Praise for *Fen, Bog & Swamp*

"This nonfiction book on a small portion of nature packs a punch."

—Cassie Gutman, *Book Riot*

"A fascinating, captivating new book by Annie Proulx that reveals the mystery and majesty of fens, bogs and swamps."

—CJ Lotz, *Garden & Gun*

"In *Fen, Bog & Swamp*, Annie Proulx shows us how to fall in love with wetlands. . . . [The book] pays the kind of artistic and emotional attention to swamps that is usually reserved for sunsets and canyons."

—Kiley Bense, *Inside Climate News*

"The Pulitzer Prize–winning author of 'Brokeback Mountain' and *The Shipping News* has written a book—really a lengthy, beautifully executed essay—on climate change. . . . In searing prose that merges history and science, Proulx wields her pen like a knife to slowly carve out what's already lost."

—James Tarmy, *Bloomberg Businessweek*

"In a way, Proulx points out, the fight to preserve wetlands is a metaphor for the global task of slowing climate change—a failure to see how small acts of destruction add up to something much larger, and a scramble to save ecosystems only when the harms to ourselves become undeniable."

—Gregory Barber, *Wired*

"In all the book's sections, the righteous anger of a climate watcher is blended with the beautiful prose this author has been writing for forty years. . . . This is a stark but beautifully written *Silent Spring*–style warning call from one of our greatest novelists."

—Steve Donoghue, *The Christian Science Monitor*

"National Book Award– and Pulitzer Prize–winning Proulx's attunement to the intricacies and vulnerabilities of nature and humankind's reckless exploitation of the living world shapes her celebrated fiction. . . . Writing with her signature vitality, precision and creativity, she crafts a galvanizing narrative out of a cavalcade of facts. . . . Proulx's concern for the future of life on earth as the planet warms is acute, while her inquiry into the watery places where peat is found balances alarm and despair with wonder and affirmation of nature's ability to rebound."

—Donna Seaman, *Booklist*

"An epic look at wetlands across the globe. The book is a gem, a profound way of seeing and feeding our souls with the richness of the fragments we have left, habitats we have an urgent duty to protect and restore."
—Catherine Cleary, *The Irish Times*

"Annie Proulx's sparkling book *Fen, Bog & Swamp* will open your eyes to humanity's reckless trashing of wetlands in the name of 'improvement.'"
—Richard Mabey, *The Telegraph*

"In accessible language, the acclaimed author of *Barkskins* and 'Brokeback Mountain' turns her perceptive eye to the destruction of the world's peatlands in *Fen, Bog & Swamp,* a short, informative history that argues for their preservation and restoration."
—*BookPage* (starred review)

"Pulitzer winner Proulx sounds the alarm on the role of earth's wetlands in the climate crisis in this stunning account. This resonant ode to a planet in peril is tough to forget."
—*Publishers Weekly* (starred review)

"With this collection of short essays about peatlands, [Proulx's] eye for folly is sharply trained on the long record of ruinous drainage 'projects.' But while there are many occasions for eco-grief in the book, there are also glimmers of hope. Fans of Proulx's fiction, even those with marginal interest in peatlands, will be intrigued by the snippets of memoir and the habits of a writer's mind that this collection reveals."
—*Library Journal*

"Remaking the world inevitably impoverishes it and us, as Proulx writes in a crescendo. . . . She provides a particularly good compact history of the draining of the fens of eastern England in an act pitting capitalists against working people. . . . An eloquent, engaged argument for the preservation of a small and damp yet essential part of the planet."
—*Kirkus Reviews*

Also by Annie Proulx

FEN, BOG & SWAMP

A Short History of Peatland Destruction
and Its Role in the Climate Crisis

Annie Proulx

SCRIBNER

New York   London   Toronto   Sydney   New Delhi

Scribner
An Imprint of Simon & Schuster, Inc.
1230 Avenue of the Americas
New York, NY 10020

First Scribner trade paperback edition June 2023

SCRIBNER and design are registered trademarks of The Gale Group, Inc., used under
license by Simon & Schuster, Inc., the publisher of this work.

For information about special discounts for bulk purchases, please contact Simon &
Schuster Special Sales at 1-866-506-1949 or business@simonandschuster.com.

The Simon & Schuster Speakers Bureau can bring authors to your live event. For
more information or to book an event, contact the Simon & Schuster Speakers Bureau
at 1-866-248-3049 or visit our website at www.simonspeakers.com.

Manufactured in the United States of America

3 5 7 9 10 8 6 4 2

Library of Congress Cataloging-in-Publication Data is available.

ISBN 978-1-9821-7335-7
ISBN 978-1-9821-7336-4 (pbk)
ISBN 978-1-9821-7337-1 (ebook)

Image Credits: page 1, Josiah Wood Whymper; page 35, Universal History Archive/
Contributor; page 73, Christine Bond @ Remco de Fouw; page 125, courtesy of
The History Museum, South Bend, Indiana

This little book is dedicated to the people of Ecuador who made their land the first country in the world to include legal rights for natural ecosystems in its constitution. The recent ruling against mining companies to protect the Andean cloud forest Los Cedros is a significant event for the world.

FEN: A fen is a peat-forming wetland that is at least partly fed by waters that have contact with mineral soils, such as rivers and streams flowing in from higher ground. Such minerotrophic water can support reeds and marsh grasses. Fen waters tend to be deep.

BOG: A bog is a peat-making wetland with a water source that is not in contact with mineral soil—rainfall. The ombrotrophic water supports sphagnum mosses. Bog waters tend to be shallower than those in fens.

SWAMP: A swamp is a minerotrophic peat-making wetland dominated by trees and shrubs. Its waters tend to be shallower than those of fens and bogs.

I followed William J. Mitsch and James G. Gosselink, *Wetlands*, 2015, 5th ed. (Wiley), for definitions and explanations of wetland processes.

Why fens, bogs and swamps?

These pages started out as a personal essay to help me understand the wetlands that are so intimately tied to the climate crisis. The literature is massive and I had to narrow down the focus to those special wetlands that form the peat that holds in the greenhouse gases CO_2 and methane—the fens, bogs and swamps and how humans have interacted with them over the centuries. The essay grew into this small book. I am not a scientist and much of the material I found was presented in specialized vocabularies which I have tried to avoid when possible. I suspect this gulf of esoteric language is an important part of the disconnection between science and ordinary readers.

There are people who are fond of tracing ideas and their connections in unlikely places and old books; I am one of them. I am easily enchanted when an odd idea or phrase looms on a page, often showing an invisible link. It is a little like a foggy summer morning when we can see beaded spiderwebs strung between stalks and stems, between tree and ground, twig and leaf. As the sun heats the earth the droplets evaporate, and the illusion that the entire world is held together by fine spider threads evaporates with it.

Annie Proulx

I.
Discursive Thoughts on Wetlands

Adventure with Curl-Crested Toucans
by Josiah Wood Whymper

I believe that whatever time you are born into shapes your perception of humankind vis-à-vis the natural world. I was born in 1935 in still-rural eastern Connecticut. Both my parents were descended from seventeenth-century settlers in North America. In 1935 they were two generations past the transition from independent farming to textile mill work and struggling with a more modern transition from the textile mills to middle-class white-collar lives, yet both families still kept chickens and a cow. My mother's family included furniture makers and artists. They all were amateur naturalists who knew the habits and habitats of birds, insects and amphibians, who could name every wildflower, every tree and the uses of its wood. They had a camp on Lake Quinebaug where my cousins and I learned to swim. They had a private reverence for mossy quiet woods, were always thrilled to see wild hawks streaming north on their spring migrations. My earliest memory is of sunlight filtering through leaves when I was put to nap under a tree. From this family in that decade I was given a glimpse of the intricate complexities of the natural world. I am anchored in that childhood time when to recognize a sassafras bush from its mitten-shaped leaves was the sense of finding a friend in the woodland fringe. I thought I knew something of the world.

As I grew older and read and traveled I learned that the 1930s were years of vile human behavior in a world that hubristically considered itself "civilized"—years of economic collapse, a global depression and mass poverty, severe and prolonged drought, gulags, strong-man leaders, intense nationalist demagoguery, ethnic atrocities, deforestation, lynching, gangsters, trafficking. In the ever-continuing name of Progress, Western countries busily raped their own and other countries of minerals, timber, fish and wildlife. They built dams and drained wetlands. It was the last decade of existence for the ivory-billed woodpecker in the southern swamps. Governments and entrepreneurs straightened and dammed river courses, smothered coastlines with rip-rap, dynamited mountains and gouged deep mines, defiled the skies. Yet my life-start in the middle of this notorious decade, part of a larger time frame that has been called the "psychozoic" now seems a time-capsule anomaly. Today I can see the period as a harbinger of the awfulness of the present. But in 1938 I was three years old and knew nothing of looming war nor murderous dictators nor exultant profiteers destroying wild places nor pandemics, insurrections or poisonously intoxicant politics.

My compassed childhood rural world was filled with novelties. One day my mother led me through a blueberry thicket to a sudden swamp where she jumped from dry land to a grass tussock, then to another, and I tried to follow. I made it to a quivering clump and looked down into the water. Something stirred up a pale cloud of mud. My mother's arm rose and fell in a descriptive arc. The next tussock was distant and its grass stalks were strung with a zigzag centerline web occupied by a yellow-and-black tigerish spider. If I jumped I would fall into the sinister water or land in the spider's arms.

4

So I bawled, and my mother carried me back to solid ground and we made our journey around the perimeter, and, where possible, into the interior of the swamp, past dead tree snags guarded by raging birds, skirting pools of water lilies whose somnolent musk no perfumer has ever duplicated. Thousands of spider strands laced stems and reeds, attached to half-sunk logs; frogs were everywhere, their pop-up eyes glaring over the edges of lily pads; unseen distant creatures splashed into hiding. It was frightening and exciting. This place, so unfamiliar and strange, was my first experience of geographical Otherness, my first thrill of entering *terra incognita*. The Polish artist-writer Bruno Schulz touched the moment when he wrote, ". . . we manage to acquire images in childhood that carry decisive meanings for us. They function like those threads in the solution around which the significance of the world crystallizes for us." For me this statement is true. I hope it was true for Schulz and that his final thought was of a golden childhood image—perhaps that lambent night described in *The Street of Crocodiles* when the child-narrator wanders the city lost in moon-daze, but more likely it was the image that haunted him through his life of "a cab, with a hood on top and lanterns blazing, emerging from a nocturnal forest" when he fell.

The memory of the swamp spider stayed with me. Years later I learned that the zigzag central line or stabilimentum in orb-weaving-spider webs may serve to reinforce the web's strength against the catastrophic blunder of a flying bird. Some believe it can attract prey as the stabilimentum is a brilliant white nonsticky silk that reflects UV waves which both birds and insects can see, but another study reported in *Behavioral Ecology* found that stabilimenta reduced the study spiders' game bags by almost 30 percent. Others offered the straw-clutching suggestion that it might have some camouflage use in disguising a large colorful spider from predatory attack. Some think spiders may be using up leftover silk, as the silk glands must be empty in order to start the process of making more. Still another idea is that female spiders make stabilimenta to attract mates, and there is some support for this. About the only idea not given credence is the original thought that a stabilimentum gives the web stability. In other words we don't know why certain spiders knit these zigzags into their webs any more than we know where or when the next catastrophe will lurch at us.

I came away from that wetland sharing my mother's pleasure in it as a place of value but spent years learning that if your delight is in contemplating landscapes and wild places the sweetness will be laced with ever-sharpening pain. In this century many people are aching with eco-grief over deforestation, the disappearance of bumblebees and ash trees, the loss of coral reefs and kelp forests; we see polar bears on a hopeless search for the firm ice of yesteryear, sage grouse and prairie chickens confronting hog farms, wind turbines and highways on their nesting grounds.

A bone-deep identification with the place of one's origin can be almost as strong in some humans as it is in animals and birds. In prehistoric times that symbiosis began before your eyes focused and continued throughout life because people were moving parts inside their landscapes. The intimacy of people close to their environment is reflected in such descriptive language as the Apache place-names Green-Rocks-Side-by-Side-Jut-Down-into-Water; Grey-Willows-Curve-Around-a-Bend; Trail-Extends-Across-Scorched-Rocks.

In human migrations our relocated emigrant ancestors had to give up allegiance to their old landscapes; the memories were emotional hawsers mooring them to their ancestral geography—birch trees in spring rain, a rocky inlet. Few today can identify with that old woman in Frank O'Connor's story "The Long Road to Ummera" who struggled like an upstream salmon to get home to die. Her longing ended with success where "The lake was like a dazzle of midges; the shafts of the sun revolving like a great millwheel poured their cascades of milky sunlight over the hills . . . and the little black cattle among the scarecrow fields."

In just my lifetime I have seen a thousand kinds of damage humans have inflicted on ecosystems and wildlife habitats as more than 60 percent of the world's rivers have been dammed and the forests massacred, ripping apart the ancient notion of the web of life. We have behaved dangerously by indulging in a global storm of greed that is fracturing biodiversity and the natural world. Since 1950 the world population has increased by nearly 200 percent. Our swelling and hungry population is spilling over, as in the title of David Quammen's 2012 *Spillover*. Quammen compares the human population explosion to an outbreak of tent caterpillars. As we cut down deep forests and convert wild places to feed lots and drained swamp to cropland we encounter other species—birds, mammals, reptiles, bacteria and viruses—alien viruses whose hosts and habitats we have severely disrupted and displaced, so that viruses such as SARS, Ebola, MERS, the cluster of "swine flus" and Covid-19 are forced to find other places, other hosts including humans.

Asian countries are hotspots for emerging new viruses partly because of regional population swell and intense deforestation, but millennia of agro-ecological intermixing in the region underlies all assumptions. The assaults on ancient forests populated by unknown micro-organisms puts humans in contact with viruses better left alone. Bats pollinate many plants and eat large amounts of harmful insects, but they also carry many viruses. When we force them out of their evolutionary homes they discover substitutes for caves in sheds and attics, in urban building recesses. These animals do not directly pass on viruses to humans. There is usually an intermediary host that humans handle or eat. For SARS in China it was the civet cat; MERS broke out in the Middle East through camels. And although both bats and pangolins were high on the suspect Covid-19 intermediary list, the pangolin has been exonerated. In the Elsevier journal *Infection, Genetics and Evolution* the authors of a study on the origin of Covid-19 concluded: "The real triggers for epidemic and pandemics are the societal organization and society-driven human/animal contacts and amplification loops provided by the modern human society, i.e., contacts, land conversion, markets, international trade, mobility, etc." In that "etc." lies our future.

Deforestation for the sake of more cropland opens another door behind which we find the pulsating bulk of animal farms, especially poultry and swine. Rob Wallace's collection of his blog essays—*Big Farms Make Big Flu*—is an aggressive probe into large-scale mono-agriculture which has replaced wetlands, grasslands and forests.

Our species is not adept at seeing slow and subtle change. We truly live in the moment. (The success of the retailer Amazon is built on this attribute.) There is a tree, we cut it down—we immediately recognize that there is a change. Yet we see a tree and we see it again a year later without noticing the new-growth tips (self-similar fractals of the tree); we see no change. We are never astonished at "the undying difference in the corner of a field." We just don't get the slow metamorphoses of the natural world because we have unplugged ourselves from it except for annual vacations which may be vehicle travel to a national park or a "nature" adventure cruise such as visiting the Galápagos or the Antarctic, where our short gawk further damages the habitat.

To observe gradual change takes years of repetitive passage through specific regions week after week, season after season, noting sprout, bloom and decay, observing the local fauna, absorbing the rise and fall of waters, looking carefully—the way all early humans lived. Henry David Thoreau (1817–1862) of Concord, Massachusetts, followed the practice of *repetitive observation*. Thoreau, periodically ill with tuberculosis flare-ups most of his adult life, tramped for miles every spring noting in his bad handwriting the dates when wild plant species flowered. His records for the years 1852–1856 were extensive. When the tuberculosis surged again, he missed listing some plants in 1857 and 1858. In 1860 he made a trip to Minnesota, his last long trip anywhere. Back in Concord he worked on editing his journals and in December 1861 he went out on a rainy day to count tree rings in a particular stump and came home soaked and freezing. He developed bronchitis which augmented the tuberculosis and by May 1862 he was too ill to leave his bed and died as the spring flowers began to bloom.

Many of his industrious Concord neighbors saw Thoreau as a ne'er-do-well, a fool who rambled the woods instead of hoeing garden rows or making the anvil ring. But some took him to heart. Aldo Leopold similarly kept records of spring flower bloom times on his Wisconsin farm for many years as did many unsung rural people attentive to the botanical seasons. Closer to Thoreau's time was Alfred Winslow Hosmer (1851–1903), also a native of Concord, a photographer and dry goods merchant who ardently admired Thoreau. Sixteen years after Thoreau's death Hosmer decided to continue the spring canvas of wild flower bloom times. He kept at it until 1902. A hundred and fifty years later the biologist Richard B. Primack and Abe Miller-Rushing followed the same observation trails for the Thoreau-Hosmer list's forty-three most common plants. They used the comparative data as hard evidence for a warming climate. Writing about the pink lady's slipper orchid and his own wildflower searches around Concord, Primack referred to Thoreau's notebooks.

> In 1853 he [Thoreau] noted the first flowering of this species on May 20, and between May 24 and May 30 in subsequent years. . . . Today, if I went looking for the first flowers of the pink lady's slipper orchid on May 20, I would be too late . . . [T]he pink lady's slipper orchid in Concord has shifted its first flowering time so that it now flowers three weeks earlier than in the past. . . . Only by comparing records taken at widely spaced intervals—Thoreau's records in the 1850s and our own observations made 160 years later—could I see the alteration in flowering times.

We have our modern-day Thoreaus in Inuit people and residents of the Marshall Islands, Miami, Siberia, Easter Island, New York City, watching the water rise, and Yakutians watching their land collapse and subside as the permafrost under their roads and fields slumps. Observant people still exist, among them the ecologist Charles Crisafulli, who began exploring the ashen landscape just two months after the Mount St. Helen's 1980 eruption. He found a plant—"parsley fern"—*Cryptogramma crispa*, a pioneer species that dares to grow in acidic scree and is a companion to volcanoes; since then Crisafulli returns to greet the plant anew every year.

Charles Wohlforth's *The Whale and the Supercomputer* examined the tensions and accords of two groups concerned with climate change—native people and scientists—in northern latitudes. Technological devices became as important as the watchers in 2018 when the Ice, Cloud and Land Elevation Satellite-2 (ICE-2) replaced an older satellite. ICE-2 gives highly precise details of the surface elevation of Antarctic ice over a vast area, showing specifically where ice is melting and where it is accumulating, with an accuracy of several inches. And yet there is a yawning gap between our past idealistic honor of the web of life, that wordless bond of interrelationships between all parts of the earth, and, in today's human-bossed world, a recipe for how a divided and impulsive species can go about "managing the entire planet as a combined physical and biological system," as a well-known champion of market-based conservation sets out as a goal. We, who cannot even manage to live peaceably together, can "manage" the whole earth? To be feared is AI, geo-engineering and an app-happy gig-rigged future controlled by Big Tech.

And yet in some places the bonds were not broken. The environmental designer Julia Watson has tracked down extraordinary examples of cooperative ecological solutions to specific regional needs. Most striking are the living root bridges and ladders made by northern India's hill tribe people who arrived there about forty thousand years ago from southeast Asia. Today they live high in the forested mountains in one of the world's rainiest places, where during the monsoon violent rivers gush through the valleys making travel between villages impossible. The region gets a staggering amount of monsoon rainfall that turns the network of villages into islands. The only bridges able to withstand the roaring inundation are those formed by training the roots of the sacred rubber fig trees across ravines, a labor that takes fifty years of work but then lasts for hundreds more. The bridges are an artful tangle of monstrously long tree roots that writhe and twist through each other as living sculpture. The Khasi's history and spiritual origin beliefs are entwined with the multigenerational design. Ladders and walkways made the same way connect the upslope villages with the farmlands at a lower elevation.

Equally ingenious are the "forest gardens" of Tanzania's Chagga people. The Chagga are banana specialists growing the fruit on the slopes of Kilimanjaro. Over generations they have modified the original forest while copying its ways and growing their gardens within the forest which surrounds each village. The diversity of their gardens is legendary. They grow more than twenty-five varieties of banana in the Kihamba groves but also cultivate more than five hundred species of plants including introduced avocado, papaya, sweet potato, mango, taro and coffee. The gardens are multilayered and interspersed within forest sections of trees, lianas, shrubs and epiphytes. This gardening style designed to fit local nature has caught the attention of people in other countries including Sri Lanka, the Pacific islands, Indonesia and Peru.

* * *

In the last decades of the twentieth century evidence of climate volatility in the American west began to catch at me as the Rocky Mountain lodgepole pine forests sickened and died from drought and the ravages of pine bark beetles. These grey tinder forests throughout the Rockies set me to writing a novel that followed a three-hundred-year-long trail of deforestation. As I write today those real-life western forests that drew my attention are burning with ferocity.

When I moved from Wyoming to the coastal waters of the Pacific Northwest I had to learn a new place where land and water worked on each other. Just to recognize the interlaced layers of estuarine habitats took time. Nothing seemed to have a solid existence; explanations of tides, sea level, bull kelp, eroding bluffs, the lives of shore birds, the forests and their understories—all came with attachments of besetting problems. I heard constant comparisons with the near past when luxuriant kelp forests made a near-shore paradise for sea life, when great swathes of the Olympic range had still not been explored, when the water roiled with pods of orcas and the whale migrations—not cruise ships—ordered Neptune's realm. I read about the extensive clam beds and giant geoducks that made the region famous; but settler-building and sewage outflows from the growing towns ended those rich days. Even the ocean was not immutable, as now its alkalinity is sharpening with acid. In some places the shoreline is augmented with riprap and breakwaters where homeowners mistakenly think they can defeat rising water. A busy railroad spoils much of the mainland shore. The snore of container ships, trash barges and tankers, speeding pleasure boats, thrashing ferries plowing their endless furrows and roving seismic vessels of oil and gas exploration damage the ocean's private silence. Underwater cacophony reigned until in the spring of 2020 during the brief Covid-19 lockdown when the waters became relatively silent; whales, whose young ones had never experienced quiet, may have been delighted. And although we are terrestrial creatures who know how to farm, this is not enough for us and we brazenly take the fish and even krill that marine animals depend on for life. The waters tremble at our chutzpah and it seems we will not change.

Nor do we wish to change. Ancient Judeo-Christian beliefs allow humans to use the rest of the world as they wish:

> And the fear of you and the dread of you shall be upon every beast of the earth, and upon every fowl of the air, upon all that moveth upon the earth, and upon all the fishes of the sea; into your hand are they delivered. Every moving thing that liveth shall be meat for you; even as the green herb I have given you all things.

Oliver Rackham wrote truthfully that the history of wetlands is the history of their destruction. Most of the world's wetlands came into being as the last ice age melted, gurgled and gushed. In ancient days fens, bogs, swamps and marine estuaries were the earth's most desirable and dependable resource places, attracting and supporting myriad species. The diversity and numbers of living creatures in springtime wetlands and overhead must have made a stupefying roar audible from afar. We wouldn't know. As humans have multiplied to a scary point of concern about the carrying capacity of the earth, wetlands were drained and dried for agriculture and housing. Today 7.8 billion people jostle for living space in a time of political ferment, a global pandemic and now a war, trying to ignore increasingly violent weather events as the climate crisis intensifies.

Peat is not a simple substance. The raw stuff is partially rotted and compressed plant material—seasonal deposits of leaves, reeds, grasses, mosses and fibers that fall and settle in the water. The water locks out oxygen, the main agent of decay. The spongy deposits build up over centuries, each bog or fen or swamp developing an individual character. Peat contains free cellulose content, high moisture and less than 60 percent carbon. Its content varies in chemical composition and macroscopic and microscopic substances. Under the top layer of a bog the "pipe peat" used for centuries for home heating has the appearance and texture of solidified chocolate pudding. It can be cut with a sharp tool. It is slightly limber when still moist and, like green wood, must be dried before it can be burned.

For thousands of years the palsa bogs characteristic of tundra regions have imprisoned carbon just by their position atop the frozen plant material we call permafrost. In some places on Alaska's North Slope the permafrost is more than two thousand feet deep. As climate heat builds, the frozen underpinnings, which now seem not all that perma, are softening. The fearful knowledge is that greenhouse gases long held below the surface are escaping, adding exponentially to the crisis by creating a positive climate feedback—i.e., increasing the warming, the same future we dread in the tropical forests. There are indications that the thaw is now irreversible. A cascade of tipping points is at hand and in October 2020 news, word that the frozen methane in the Arctic Ocean was beginning to escape should have sent chills down our spines. At the same time we try and extract crumbs of hope from each catastrophic manifestation of change. In Yakutia, where the earth's largest thermokarst—the "Batagai Depression"—a kilometer in length, 100 meters deep and still growing rapidly, is also the site for scientists who are gathering samples of preserved and frozen soils up to 200,000 years old. Using molecular genetic and microbiological procedures they hope to extract bacteria and viruses from the soils that could lead to new antibiotics. But for the third straight year terrible fires are devouring the forests of Yakutia. The smoke drifts out into the Pacific and people on ships and in Alaska and along the North American west coast cough and gasp in the pungent brown haze.

Since the fifteenth century when feudalism began to give way to nation-states, Western capitalism and imperialism, we have heard that peatlands are worthless because that same land *drained* is valuable for agriculture. We are now in the embarrassing position of having to relearn the importance of these strange places that are 95 percent water but fibrous enough to stand on. Climate, weather, season, earth movement, wet and dry environments are all flexible, all give-and-take, and these jostling, shifting processes are only briefly impressed by levees, dams, drains, dikes, culverts. Water is the ultimate flexibility as Fela Kuti sang in "Water No Get Enemy." It will always win. Or will it? Some researchers think that the next fifty years will see humankind take up all the remaining land on earth for agriculture and use every drop of fresh water. And then what?

Europeans, especially in northern Germany and Ireland, have harvested peat by hand for thousands of years with specialized tools, adapted in Ireland as the slane and turf barrow before the invention of heavy peat-cutting machinery. But now a great change is under way; some people are rewatering and restoring old peatlands. Ireland, with plenty of bogs and few oil wells, and thus more dependent on peat than most, has been struggling with difficult decarbonization goals since a study found that in one year each hectare of drained and stripped peatland gassed out 2.1 tons of carbon. Britain, freed by Brexit from old EU rules on farming subsidies to promote intense agricultural production, is considering more projects for a category called "the Public Good," which includes wetland restoration. The richly encouraging books *The Shepherd's Life* and *Pastoral Song* by the British farmer James Rebanks give hope for a turn to thoughtfully ecological agriculture.

The world's largest peatlands are Canada's Hudson Bay Lowlands, Russia's Great Vasyugan Mire, the Mayo Boglands, America's Okefenokee National Wildlife Refuge, Indonesia's peat forests, the Magellanic Tundra Complex of Patagonia, Tierra del Fuego and the Falkland Islands, the marshes of Mesopotamia and the Central Congo Basin's Cuvette Centrale. Indonesia's richly complex wooded peatlands where entrepreneurs log, burn and plow to make palm oil plantations are one of the saddest examples of great biological loss. In shops and stores I read labels and when I find bars of soap made with palm oil I get a mental image of a ravaged forest. I do not buy that soap.

While some people living in peatland countries may be looking at their watered topography with new respect, some are looking with fear, as the residents of the Yakutia region of eastern Siberia, one of the fastest-warming places anywhere. Here, and in the Great Vasyugan Mire in western Siberia, the permafrost is going fast—the cover is being lifted off the pot.

The permafrost in Yakutia, a major agricultural region, exists as great lobes of ice—a kind of very thick ice called *yedoma*—embedded in the soil like garlic slivers inserted in a roast. Up on top, land that was recently crop acreage, cattle and reindeer pasture is swimming in meltwater. Great chasms open up, roadways tilt and slump. Weird craters, possibly caused by underground bursts of methane gas, gape wide. The rivers bulge with water and flood the fields. Yakutians, intensely sensitive to the subtleties of the natural world after deep generations on the land, now say they no longer understand the place. They do know that they can no longer make a living here and are moving out of the known rural world of their fathers into the wilderness of cities. The only ones who are happy are hunters of the mastodon tusks exposed for the first time in millennia, the corpses emitting a great stench as they rot—for the hunters it is the stench of money as long as Chinese folk medicine believers keep buying. That is the frightening side of peatland's ability to hold in huge amounts of carbon dioxide: rip or burn the cover off and it is in your face.

In aggregate the world's peatlands resemble a book of wallpaper samples, each with its own design and character—some little more than water and reeds, others luxuriously diverse landscapes of colors we urban moderns never knew existed, silent sepia water, brilliant mosses, pale lichen, sundews like spilled water drops. And always they are in achingly slow motion that we cannot discern unless we keep measurement records—you can stand for a year and watch though you won't see a saltwater marsh silt up and become a fen. And always these places are under assault.

*   *   *

Wetlands are classified by the values of what-use-are-they-to-humans. They are categorized as geography, topography, chemistry and hydrology. Ecologists use a different measuring stick; they are interested in the different ways wetlands fit into the natural world's mesh of existence. In recent decades the ecologists have had a seat near, if not at, the classification table. A cool and dispassionate scientific approach to wetland classification in these times often masks painful emotions and grief when confronting environmental loss and destruction. These feelings are as much an occupational hazard for environmental scientists as for people working in disaster relief and health care fields. The painful stresses are steadfastly ignored by most of us so there is little amelioration.

The various wetlands schema fill reference books. The United States Fish and Wildlife Service uses the Cowardin Classification of Wetlands and Deepwater Habitats of the United States. It is a system that identifies five wetland types: marine, tidal, lacustrine, palustrine, riverine. The U.S. Army Corps of Engineers additionally uses a Hydrogeomorphic Classification for Wetlands. Some states have their own systems. Washington State has an extreme range of habitats from its Pacific shoreline to the easterly dry Scablands. It lists four categories of wetland ranging from the *heavily disturbed* to the *rare and irreplaceable*. Alaska, finding the standard grades "insufficient to meet the needs of land managers in Alaska," devised the Cook Inlet Classification to understand the oligotrophic peatlands in the face of the rapidly growing population of Cook Inlet Basin. Different countries have different classifications. On a flight across Canada (or looking at a topographic map of that country), you can see countless blue speckles of glacier meltwater. (You can also see the ravaged tar sands peatland region as you fly over Alberta.) Detailed descriptions of these waters are the stuff of the Canadian Wetland Classification System (CWCS), which purports to recognize each blue dot's ecological details. In practice, for most countries local mapping and characterizations are beginning to augment the older one-size-pretty-much-fits-all wetlands definitions.

Most classifications now include a scale of protections for wetlands. The Cuvette Centrale, the size of the United Kingdom, discovered only in 2012, is protected by the Brazzaville Declaration of 2018, a consortium of the Democratic Republic of the Congo, the Republic of the Congo and Indonesia and several Global Peatlands Initiative members. But we can see in the examples of Brazil and the United States and other countries how legislation can be trimmed and fitted to the desires of entrepreneurial humans rather than the health of the natural world. The Global Peatlands Initiative, led by the UN Environment Programme, exists to help peatland countries save or restore these vital wetlands which cover about 3 percent of global land area. The legally binding Paris Accord was signed in 2015 by all of the world's countries except Nicaragua and Syria with a goal of keeping the earth's temperature rise below 2° C. In several ways it was obsolete before the ink was dry—it did not discuss peatlands beyond recommending that each country take steps to preserve the wetlands it had. In 2016 the climate-change denier Trump pulled the United States out of the agreement. President Biden rejoined, but confronted with the manic drive to profit from the earth's shrinking resources we are left with the nagging question "Is this enough to save the habitable earth?"

When the facts of our self-inflicted climate crisis began to percolate into public discourse in the late twentieth century the Amazon forest was considered, after the ocean, the great storehouse for CO_2. (Overlooked for decades, coastal salt marshes were thought not to hold in CO_2 but in 2020 Canadian scientists discovered that "blue carbon sinks" like mangrove forests and salt marshes embrace half the CO_2 in ocean sediments.) Climate-change worriers pinned their hopes on the vast tropical forests. But a few scientists recognized the carbon-holding capability of peatlands and their voices grew more insistent as they pointed out that not only does peat absorb and sequester huge amounts of toxic gases, it covers 3 percent of the world's surface, larger than all the earth's rain forests together. Yet even those wetlands cannot replace the Amazon as a climate moderator. At that time it was unthinkable that such a great and enduring ecosystem as the Amazon could ever burn. We have learned a bitter truth—deforestation and fire are its new enemies.

The Amazon basin has been a gaspingly diverse region that nurtured the earth for 55 million years until Western humans entered it in the sixteenth century, chopping into the varying indigenous populations. The science journalist Marco Lambertini calls it the "richest botany in the world." He stresses the diversity: "For example, 300 different species of trees can be found in just 2 square km (500 acres) in Brazil, compared to a few dozen species in an extensive woods in the temperate or boreal zones." This intricately knotted zone

21

helped regulate earth's climate through water transpiration and cloud formation. The tropical forest drove the circulation of global atmospheres as well as stored carbon and biomass. The region's other ecological parts worked with it in a many-layered call-and-response manner.

Outsiders tend to regard only the tropical rain forest as the Amazon, but the entire basin-filling assortment of ancient bits and pieces of land and animals that originated in other places (Africa, North America, Australia, Southeast Asia, Antarctica) over millions of years have come together in jaw-dropping variations of existence that make up an unmanageably diverse mix of ecosystems, so complex that the human brain cannot fully grasp its intricate workings.

> There are lake and river communities, inundated flood-plains called *varezas* [*várzeas*], and lands beyond the floodplains called terra firme . . . topographic and edaphic variations within the basin that support basins of grass-land, savannah, and dry forest on promontories and sandy soil, and the surrounding cerrado and caatingas extend into the basin varying distances from the surrounding margins. These different habitats and their biota are important because they allow for rapid response of the veg-etation to environmental change.

It is impossible to overemphasize the importance of that final sentence—many species can survive temporary adverse events if there are reachable areas (refugia) that satisfy their basic needs. The paleobotanist Alan Graham has constructed a mind-bogglingly detailed chart of the natural and human-caused interlocking forces that affect climate: thermohaline cells, Milankovitch variations, sea level, ocean basin volume, CO_2, methane, atmospheric circulation, rainshadow, El Niño, albedo, volcanism, vegetational history, rates of erosion, orogeny, vicariance, new climatic zones, oceanic circulation, land bridges, terranes, plate movement and tectonics, seasonality, degassing, the solar constant and even the shapes of continents. The human impact is everywhere. Although there has been a population increase in Brazil's endangered golden lion tamarin, in May 2018 conservationists reported the first yellow fever death of a tamarin, a disease transferred from humans to the tamarin by mosquito vectors. In the following year the sickness wiped out more than 30 percent of the animals. Adding this loss to ongoing deforestation, fires and wetland drainage projects the current Brazilian government plans to reconstruct and extend the major highway BR-319 across the Amazon continuing into "protected" forest.

The Amazon has persisted for millions of years because of its great size and because its diverse ecosystems and habitats have functioned as plant and animal refugia in times of painful climate disturbance. But as sections are deforested, burned, highwayed or plowed for soybean cultivation the refugia disappear and the whole is damaged; without its missing parts the Amazon's flexibility stiffens. The twenty-first century is apparently too much for it; it is suffering massive attacks on three fronts, especially in Brazil: both authorized and criminal deforestation by cutting and fire; degradation by roads, cattle ranching, large-scale agriculture; the increasing drought and heat of the worldwide climate crisis.

When I began to think about writing this essay two years ago the once self-sustaining Amazon was teetering on the point of a flip from closed canopy rain forest to the mix of grasslands and trees known as a savanna. On July 15, 2021, *The Guardian*'s environment editor covered the release of the article "Amazonia as a Carbon Source Linked to Deforestation and Climate Change." The study gave us bad news: that ten years of measuring the Amazon's CO_2 output shows that the CO_2 levels have increased as fires and apparently unstoppable deforestation continue. The Amazon is now emitting more CO_2 than it sequesters. It is not yet clear that the shift is irreversible but the changing conditions—less rain, fire, deforestation—indicate the worst. Once the process begins it becomes increasingly difficult for the rain forest to return. Open grassy savanna would be even more prone to fires which inevitably keep the area dry and open. We need rain forests for their unparalleled diversity of species and because they absorb much more CO_2 than can any savanna.

It is a peculiarity that the small group of Spanish military adventurer-explorers in the Americas following Columbus were "a tight circle of like-minded individuals, many of them from harsh boyhoods in the Extremadura, some even related by blood." Balboa, Cortés, Alvarado, Pizarro, Valdivia, Orellana and De Soto "all hailed from within fifty miles of one another in the impoverished, drought-ridden, impossibly torrid highlands of western Spain." These hard men with nothing to lose—"predators hoping for easy plunder"— reorganized the known earth and imposed their grasping values on vast populations. Who can ever forget Barry Lopez's retelling of Cortés's deliberate and cruel burning of the aviaries of Mexico City which the explorer had previously described as "the most beautiful city in the world"? These were the louts many historians and politicians set up in our minds as role models. These men are a type now more numerous and harmful romping along under the flag of global capitalist business.

Francisco de Orellana, a friend and likely a cousin of the Pizarro brothers Francisco, Juan and Gonzalo, was a soldier with Gonzalo Pizarro in Quito in 1541. Gonzalo ordered Orellana to explore the Coca River in a brig, travel to its end and then return. But it did not happen that way and no one has been able to separate truth from lies in the centuries since. Where the Coca met the Napo River Orellana did not turn back but formed a new expedition to continue downstream. Did the decision come from necessity (as he claimed) or a desire to follow the heart-beating search for a city of gold? Before they pushed away from the Coca, Orellana got his men to sign a document saying the river current was too strong to allow them to return to Pizarro. So they continued on, suffering insect bites and increasing hunger until they were forced to eat their leather shoes. It was a hellish trip. The Amazon historian John Hemming remarks: "It was, and still is, extraordinary how Europeans never learned to live sustainably in the world's most diverse ecosystem."

It was June 1542 when Orellana and his men entered the Amazon and rode the great river all the way to the sea, the first Europeans to travel its length. On the journey they passed a settlement they called "Pottery Village" for the exquisitely beautiful ceramics the residents made, today known as the Guarita style, equal in quality to the masterworks of the Peruvian Shipibo and the famous vases of the ancient Greek world.

Back again in Spain Pizarro denounced Orellana, saying he had deliberately disobeyed orders. But Orellana produced his crew-signed documents stating they had been unable to paddle against the powerful river current. The Pizarros were no longer in the royal favor and the Spanish king (Charles V aka Charles I) turned away and granted Orellana's request for a commission to rule the territory he had "explored." The Spanish invaders were sick with the lust for gold, that "uncontrollable, devastating fever," a condition which once infected Alvaro Mutis's roving sailor Maqroll. The Spanish and Portuguese were the shock troops of conscienceless European explorers who rushed in eager for the aureate gleam, bright feathers, slaves, power and position, crusading Christians willing to commit any atrocity to get those rewards. But for neither the Pizarros nor Orellana did the dream come true. On his triumphant return to the Amazon Orellana and his new party, including his wife, got irretrievably lost in the countless islands at the river mouth and never found the main flow. Orellana perished, his wife said, of "grief over his lost men and from illness."

John Hemming estimates that at the time of Orellana's voyage at least four or five million humans in hundreds of groups lived in lowland Amazonia. Like the countless plant and animal species the hundreds of indigenous groups were then strikingly diverse, but by the 1980s the wasted population had crumbled to a few hundred thousand. The forest tribes, like the plants and animals, continue to suffer annihilation by the forces of income-hungry governments and ambitious agriculturists and loggers. There exists a depressingly long list of the names of indigenous Amazonian conservationists who have tried to protect their home geography and were murdered. The region still contains half the world's tropical forest, still leads all other earthly places in biodiversity, but for how long?

To understand some of what has disappeared from that world I look to the past. Two young men, the naturalist Henry Walter Bates, studying animal mimicry in support of Darwin's theory of natural selection, and Alfred Russel Wallace, traveled to Pará, Brazil, in 1848. They entered the vast forest which then was a place of high rainfall and rare fire. Its position on the equator meant tremendous solar energy pouring down in perpendicular rays, but in the humid gloom of the forest floor they had to squint up through giant trees to glimpse the blasting sunlight at the treetops, the canopy world of specialized birds, insects, fruits and flowers so different from the ground-level forest it might have been on another planet. Bates wrote of what they saw on that first entry into the forest:

> . . . the tree trunks were everywhere linked together by sipos; the woody, flexible stems of climbing and creeping trees, whose foliage is far above, mingled with that of the taller independent trees. Some were twisted in strands like cables, others had thick stems contorted in every variety of shape, entwining snake-like round the tree trunks, or forming gigantic loops and coils . . .

Today you are likely to see a giant truck grinding through deep vermillion mud while hauling out the great logs of the vanquished and smoldering forests.

Images of the Amazon more familiar than Bates's tangled loops or the thrill-packed television "documentary" series *Dicing with Death* that shows poor indigenous people risking their lives to cut down the forests are the aerial photographs of deforestation—the Amazon is difficult to photograph, as the humid area is usually obscured by cloud (or smoke)—to give an idea of its expanse. The photos show what looks like an endless broccoli farm. But at ground level it is different. Although I cannot see the Amazon as it existed in Bates's or the *conquistadores'* time, and perhaps cannot bear to see it in Bolsonaro's time, I could gaze hungrily on the 1819 painting *Forêt vierge du Brésil* by Charles-Othon-Frédéric-Jean-Baptiste, Compte de Clarac, a work of art that shaped European ideas of tropical forest, even as reports of the handsome, healthy inhabitants fed Rousseau's idea of the "noble savage."

The painter de Clarac managed to attach himself to the Duke of Luxembourg, who was on a mission to Brazil as Louis XVIII's ambassador. In Brazil on the bank of the Paraíba do Sul river the artist made several sketches of the unbelievable forest in front of him; back in France he incorporated his sketches into a large watercolor once owned by the naturalist Alexander von Humboldt but now in the Louvre.

I look into heavy brown gloom and see immense two-hundred-foot-tall trees with great buttressed wings, ferns the size of mature apple trees, upward twists of swollen coils of lianas, epiphytes, pinnate fronds, aerial roots, hollowed giants and a clump of rickety thin saplings starved for light. There is no wind. A banner of thinned sunlight illuminates a family crossing the river on a fallen log and a native hunter—tiny humans too small to leave a mark on the ageless forest. I can almost smell the mix of decay, orchidaceous perfume, the mist rising off the frothing Paraíba. Today nothing of de Clarac's painted forest remains. Plagued by drought and water allocation mismanagement the Paraíba is dammed, polluted, some of its fish extinct, home to invasives and unable to supply enough water for pressing industrial demands.

In 2019, almost in concert with each other, runaway forest fires too vast and violent for humans to stop charred the Amazon forest, Australia's parched Queensland, the Siberian peatlands and California. Each year since then the fires increase in number and ferocity. The Amazon fires were caused by a toxic mix of weather, malignant political and business forces, illegal miners and land-hungry farmers as well as the foreign investment interests that have preyed on Latin America since the *conquistadores* arrived. That this forested landscape, historically drenched with rains and veined with massive rivers, could burn was shocking. The Russian fires of 2019 burned 13.1 million hectares, including 4.5 million hectares of Siberian taiga. The conflagration was downplayed in Russia as naturally occurring forest fires despite reports of arson to hide illegal logging operations. Some believe the fires spurred permafrost melting. Once again in 2020 fires raged in Siberia. Australia's catastrophic conflagrations may become the stuff of legend—but for our time they were terribly real with more than 15 million acres burned, an estimated one billion animals killed, including not only the iconic koalas and kangaroos, but emu and other birds, bats, frogs, insects, fish, reptiles and people. The 2020 global fires included that paradise of wildlife, the Pantanal, crucial habitat for stunning concentrations of rare wildlife, including blue hyacinth macaws, jaguars and endangered giant otters. There has never been a year like 2020 for the Pantanal—25 percent of the fabled Heritage Site burned. In 2019 and 2020 the American western states, especially California, struggled again with big hot fires. Now in 2021 the peatlands are ablaze

earlier than ever. In 2020 the Arctic blanket peat was suffering from aptly named zombie fires, fires that seem to be dead
but which smolder invisibly underground through the winter, then leap back to life with warmer spring temperatures,
releasing huge amounts of carbon. There is an account of a
nineteenth-century traveler in Siberia who heard

> . . . the ground beginning to sound hollow as a drum
> under the hooves of the horses pulling the sledge. The
> driver explains that subterranean fires eat their way to
> the surface, causing subsidences, great hollows in the
> burnt earth, so that there is always a danger of a horse
> breaking the crust and sinking into the fire.

The traveler was that intrepid and scandalous English
missionary-nurse, Kate Marsden, whose 1891 book *On Sledge
and Horseback to Outcast Siberian Lepers* earned her an explorer's reputation for courage, stamina and grit as well as enemy
detractors. She added to our knowledge of the damage done
by zombie fires by explaining that the fires smoldered underground through winter and summer, burning out the tree
roots so that when summer storms with rain and wind hit,
great patches of rootless trees fall over en masse.

I remembered then the purple-red roads in northeast
Wyoming, most beautiful in the rain and somehow familiar. They were built, I was told, from the clinkers of ancient
burned-out coal beds in underground zombie fires. Then I
understood why the color was familiar. In the 1940s my parents heated our house with coal, and my father raked out
the unburnable mineral clinkers which he carried out to
our clinker heap in the backyard. The clinkers were the same

purple-red—the mineral hematite—of those western roads. Today nothing is so simple as burning coal and dumping out a bucket of clinker. Tiny bits of information appear at times like fireflies in the summer nights and we humans tear our hair trying to gather enough pertinent data to understand the simultaneous effects of climate change, the vast burning of fires above and below, the interlocking risks, the changing weather, smoke paths, the twisting winds, the play of storms, driving on coal clinkers and a constant, low-grade guilt.

As horrific and destructive as the fires of 2019 and 2020 were, there was another damage beyond incinerated trees and understory, beyond millions of animals and birds roasted alive, beyond the poisonous smoke that made breathing creatures retch and strangle and die. The extra helping of trouble from forest fires is a heavy deposit of soot on melting ice as happened in Greenland from the boreal fires in Siberia. The soot reduced reflectivity, absorbed heat and pushed the melt rate higher.

I cannot guess, after the fires, how much of the Amazon and Pantanal can regenerate and how much will be made into cattle pastures and cropland and soccer stadiums, but judging from past events recovery does not seem likely. How much of the still-smoldering Siberian taiga can recover? Can the forests of the American west come back? We know with surety that there are tipping points when badly savaged forests are unable to regenerate. One recently published thirty-year study that tracked the amount of carbon absorbed by 300,000 trees in the Amazon and African tropical forests found that today the trees are taking up 30 percent less carbon than when the study started in the 1990s. In the 1990s the trees absorbed 17 percent of human-produced CO_2 but in the second decade of this century they took in only 6 percent. The projection was that by the mid-2030s the tropic forests would no longer absorb CO_2 but start releasing it. But that tipping point has already arrived, and if climate change seems to be accelerating it is apparently because the Amazonian contribution of ejected CO_2 is pushing the process.

It is no wonder that some see the Covid-19 pandemic as a sign that our destruction of the natural world has opened the way for once-sequestered diseases and viruses whose natural habitats and hosts we ourselves have now become. Before the last wetlands disappear I wanted to know more about this world we are losing. What was a world of fens, bogs and swamps and what meaning did these peatlands have, not only for humans but for all other life on earth?

In the few years I lived in Port Townsend, Washington, I often walked along the North Beach bluff at Fort Worden, which was, in 1900, an important link in west coast sea fortifications, and is today a gold star state park with lively art and science programs. On a walk with a geologist friend she pointed out a dark layer of peat at the base of the feeder bluff. Close up I looked at grey squashed layers of plant material. To me the protruding strap-like material looked like pressed reeds. It was woody, damp and crumbly, unlike the dried bricks of Irish peat bogs. In the shale-like layers I hoped to see the rare mineral vivianite, a hydrated iron phosphate said to be intensely blue when exposed to air. Ötzi, the mummified Alpine hunter who emerged from the ice in 1991, was dotted with vivianite, a mineral that forms through contact with organic material, especially iron. When in 1979 the Alaskan archaeologist Dale Guthrie unearthed "Blue Babe," a now-extinct Pleistocene steppe bison, the animal was encrusted with vivianite. Many of the ancient bodies found in northwest Europe's bogs showed vivianite. Vermeer used the mineral for the background of the carpet in *The Procuress*. If the color was an intense blue when fresh, it has faded over the years to a dull grey-green—we must imagine the vivid blue when the painting was still on his easel. In the local peat I found only minuscule white flecks. A year later when a large chunk of the bed fell onto the beach I finally saw the electric blue scattered through the freshly exposed layers of ancient peat.

The flattened stems in the peat bed at the base of the bluffs of Admiralty Inlet stayed on my mind. At first I imagined they were sedge that had once grown in a squelching fen. I am a follower of the historian Fernand Braudel, and believe with him that we can follow threads into the past by looking at "climatology, demography, geology, and oceanology, and . . . the effects of events that occur so slowly as to be imperceptible to those who experience them." In the Puget Sound region the frigid and curling Vashon tongue of the Ice Age started melting about 16,900 years ago. There were hundreds of centuries of pulsing on-again-off-again shove and drag as the ice lobe crawled south, retreated, advanced again through glacial and warmer interglacial periods. In the warm interglacial periods forests and woodlands grew until the next press of ice crushed them, until finally the grinding assault was over and in its place were rivers and braided streams of meltwater. These wet conveyer belts rearranged layer upon layer of soil and rock and mammoth deposits that over thousands of years built up the sandy bluff. The interesting peat bed had likely not been a sedge fen, but a woodland swamp. I felt that shiver of recognition of the constant and deep currents of endless change, of distant rainfall becoming maniacal flood, of sucking drought, of call and response in every fiber, grain and atom of every thing.

2.

The English Fens

"Fen. A peat-accumulating wetland that receives
some drainage from surrounding mineral soil and
usually supports marshlike vegetation."
—Mitsch and Gosselink

The Rivers of the Wash: A Bit of Fen

Alexander Pope's eighteenth-century idea of *genius loci* appears in his *Moral Essays* as advice to landscape designers to keep in mind the "genius" or spirit of a natural place. This still has meaning when considering fens, bogs and swamps. I wanted to understand why we are suffering disasters of climate change, deforestation, drought and flood, runaway fires, viral pandemics, headaches, depression and political unrest, and if the loss of natural wetlands was a key element in this unraveling I wanted to know something of how they formed, how they changed and why, when humans ignored the *genius loci*, they disappeared. I wanted to know how humans interacted with wetlands in the past and the present. The old image of an infinitely complex "web of life" holding the world together is still serviceable, but the web's self-healing gossamer has been torn by humans in so many places it no longer functions. I found the story of the English fens was a story of the tearing-apart.

Peat builds up in fens, bogs and swamps and all of these stages are constantly in flux. The wetlands are related through a successional gradation that can end as a field of soy beans or a parking lot. After the marine marsh the fen appears, defined as a "peat-accumulating wetland that receives some drainage from surrounding mineral soil and usually supports some marshlike vegetation." In their day the English fenlands were the best-known of their kind. Once as much as 6 percent of Great Britain was wetland, most extensively along the east coast of Norfolk, Lincolnshire and Cambridgeshire known simply as "the Fenlands." They were a watery mix of fresh water from rivers and streams, sea water and the land around the Wash where the Ouse River entered the North Sea—fen water deep enough for eels and fish and boats. There were hills and islands of dry land high enough for houses and gardens.

My interest intensified when I read Eric Ash's *The Draining of the Fens*, his study of the sixteenth- and seventeenth-century drainage projects, an exploration of how vested interests and political clout reshaped a large wetland region and nurtured the nation-state at the cost of its ancient ecology. The generous character and extent of these fens seemed beyond anything I had known of England's topography. Ash frequently referenced *The Fenland, Past and Present* by the meteorologist Samuel H. Miller and the geologist Sydney B. J. Skertchly, published in 1878, and one of the first attempts at a descriptive history of the fens.

When my print-on-demand copy of *The Fenland* arrived I scanned the table of contents, the list of illustrations, and then the Subscriber's List (a publisher's old way of making sure there was enough money in the kitty to pay for printing a large book unlikely to be a hot seller) in hope of finding a luminary of the day, perhaps Darwin, who was then still alive. He was not listed as a subscriber but turned up in an appendix as a contributor of unspecified information—possibly on an invasive North American water weed.

Toward the end of the list I saw the name Alfred Lord Tennyson, followed by a black blotch. An inky pen had obliterated the honorific "Poet Laureate" and above the scratched-out words had written "Plagiarist and Ass" in spiky anonymous strokes of his pen. There was more scribbling—much more—throughout the book. Since I knew nothing of the defacer but his strong masculine handwriting I thought of him as the Acerbic Hand.

Where Miller and Skertchly quoted several verses of "Camelot" he wrote "ROT" in two-inch-tall letters. Where the authors referred to "the poet" the Acerbic Hand altered it to "so-called poet." But when the editors added "a few passages in the writings of the Poet Laureate, whose style bears the impress of Fen scenery and colouring . . ." the Acerbic Hand summed up his opinion of "The Lady of Shallot" with the marginal notation: "there never was a more effeminate rotten minded milquetoast . . . idiot . . . than plagiarist Tennyson."

To give him his due, the Acerbic Hand was knowledgeable on the Fenlands, commenting on regional boundaries, history, etymology, the stones of Ely Cathedral, tillage acreage. When I saw that he had put check marks beside the names of more than 250 butterflies I wondered if he had been a lepidopterist or at least a collector who, with happy memories, was ticking off specimens he had captured. It is interesting that the only insects listed in Miller and Skertchly's compendium are butterflies. There is no mention of mosquitoes, whose link to malaria was not yet known.

Christine Cheater, in her essay "Collectors of Nature's Curiosities," described how in England "By the mid-nineteenth century, the pursuit of nature had become a craze." One of

the most extreme examples is that of a wealthy Yorkshire law-
yer who had a penchant for beautiful bird eggs. He fixated
on the eggs of an unlucky guillemot who nested in the limestone cliffs, and paid someone to collect the eggs from her
nest every year, year after year. This bird did not pass along
her genes for beautiful eggs. Some social scientists have seen
the nineteenth-century British passion for butterfly collecting as a reflection of the drive to possess an empire. Others
see it as nascent scientific methodology. Cheater notes wryly
that studies of the period's collecting fever ascribe it to "the
expansion of the British Empire, romanticism, nationalism,
the birth of the natural history museum, interior decorating
and the activities of commercial enterprises."

Among the improbably diverse group of collectors there
was also the emergent scientist. When he was young Darwin
collected beetles and in his autobiography he declared his
love:

> I give a proof of my zeal: one day, on tearing off some
> old bark, I saw two rare beetles, and seized one in each
> hand; then I saw a third and new kind which I could
> not bear to lose, so that I popped the one which I held
> in my right hand into my mouth. Alas! It ejected some
> intensely acrid fluid, which burnt my tongue so that I
> was forced to spit the beetle out, which was lost, as was
> the third one.

The depth of a collector's passion is exemplified in Alfred Russel Wallace's first glimpse of *Ornithoptera croesus*, aka
the "golden birdwing," a denizen of damp Indonesian forests.
"My heart began to beat violently, the blood rushed to my

head, and I felt much more like fainting than I have done when in apprehension of immediate death, I had a headache for the rest of the day."Vladimir Nabokov described his personal fixation:

"Few things indeed have I known in the way of emotion or appetite, ambition or achievement, that could surpass in richness and strength the excitement of entomological exploration."

There may be yet another reason for the collectors' attraction to the iridescent and brightly colored insects. The English were sensitive to vivid hues after the accidental 1856 discovery of aniline dyes. A young genius, William Perkin, loved chemistry. When the precocious youth was eighteen he began to search for a synthetic way to make quinine, the only known treatment for bone-shaking malarial diseases. When he tried coal tar as a base he did not get quinine but only a thick brown sludge. Cleaning out his flasks with alcohol he was startled to see the sludge change to an eye-searing magenta. It was the first synthetic dye and he named it Mauveine—a wonderful name perhaps for the heroine of a period novel. The brilliant and intensely saturated magentas, heliotropes, flames, buttercups and malachite greens captivated the fashion world nourished by Britain's dominant textile industry. It took but a turn of the head and open eyes to start collecting brilliant butterflies.

Even in the last quarter of the nineteenth century when the Acerbic Hand was marking up his copy of *Fenland*, England was home for a large number of native butterflies, with the fens particularly rich hunting ground.

There was mile after mile of short-cropped downland sheepwalks alive with Adonis and Chalkhill Blues; every parish contained ancient woods managed as coppice with a regular supply of warm, sheltered, flower-filled glades for Fritillaries, White Admirals and Hairstreaks. There were no insecticides, no artificial fertilizers, filling-stations or out-of-town supermarkets. Major roads and highways only infrequently criss-crossed the landscape. . . . Pollution was confined to the larger cities. The patchwork countryside, which since time immemorial had supplied hay and pasture for horses and farm animals, timber and fuel-wood, corn and orchard was also a paradise for insects.

All that changed with fen drainage. Such rarities as the large copper, whose larval food was the great water dock (*Rumex hydrolapathum*), and the Manchester argus, when deprived of the heath and cottongrass it needed, disappeared. Whittlesey Fen was the large copper's habitat and when that fen was drained in 1851 the insect was doomed. The swallowtail, which depended on milk parsley plants in sedge fields, was another victim. The piecemeal drainage of the fens over centuries by various "projectors" made fatal problems of fragmentation. No longer were the fens' many distinct habitats parts of a water-connected whole. The ecologist Ian Rotherham explains how species were lost when the fens were drained.

> . . . in an isolated pocket of habitat, literally an island in an unfavorable sea of farmland, the results are disastrous. Here we see evolution in action. The conditions most

favor individuals that do not disperse widely, but which remain in their original breeding site. This may help them survive but it also encourages inbreeding and discourages any wider dispersal to new areas. This appears to be what happened to the swallowtail in the Fens.

When Whittlesey Fen was drained Charles William Dale, the collector son of the famous lepidopterist J. C. Dale, observed: "What was once the home of many a rare bird and insect became first a dry surface of hardened mud, cracked by the sun's heat into multitudinous fissures, and now scarce yields to any land in England, in the weight of its golden harvest."

The protestors were the fen people—poor, and considered of low social class. They fought against the drainage "projectors" but were worn down and overcome after centuries of protest. In the seventeenth century the Crown laid its heavy weight on the side of drainage projects. Laws, patents, "rights"—all the machinery of bureaucracy pressed on the fen people. The drainage movement had changed over the years from a few protested local projects to a broad regional movement for drainage that was lamented but accepted.

Here and throughout Skertchly's *Fenland* we see the Victorian mindset—while many English people during the final years of draining the fens expressed deeply felt sorrow and loss, in the same breath they praised the crops of wheat and maize that replaced the wild wetlands. I am reminded of the comment of one of the men opening the American west who opined that the defeat of the Indian was "essential albeit tragic." This too is part of the human psyche—a burning sense of irrevocable loss yoked to a fatalistic acceptance of "progress" and "improvement" and the hubristic idea that "now"—the time in which *we* live—is superior to all previous times. The proofs given are usually technological "improvements."

Before the Fens There Was Doggerland

At the height of the Ice Age twenty thousand years ago Britain was not an island but a shoreline fringe on the huge landmass of Europe and Asia. This land mass included a swath that extended from the proto-England to the Netherlands on undulating plains now under the North Sea. This was the lost country of Doggerland, a sizeable 180,000 square kilometers of rolling hills.

East Anglia became a major part of the fens, but before the days of rising waters it was low-lying forest of pine, ash and oak inhabited by Paleolithic people. There may have been even earlier habitants—the species known as *Homo antecessor* from whom Neanderthal and *H. sapiens* descended. Half a dozen brawny rivers carried water down from the chalk uplands, across the lowlands to empty into the North Sea. (Later silt deposits at the mouth of the Ouse formed the distinctive rectangular shape of the Wash that shows on maps.) The residents of Doggerland likely enjoyed, as did Finn Mac Cool, the "satisfying ululation [that] is the contending of a river with the sea." As the ages rolled, the gushing rivers and the rising waters of the North Sea alternated, leaving layered deposits that became silt peat and freshwater peat across these lowlands.

The people who lived in Doggerland's woods and hunted in the hills were at home on its extensive plains. But far to the north the glacial melt was on and around 6500 BC the North

Sea began creeping inexorably across Doggerland. This was climate change with a vengeance. The sea pressed in, crested hillocks and rises, flooded low ground. There were fogs and the cries of frogs and waterbirds, the sky-darkening skeins of migratory flights, the stealthy upward creep of chill seawater.

Surprisingly, the people "didn't leave as sea levels rose; they changed their diet." Meat hunting gave way to fishing as forests and plains became fens and rivers. Logically it would seem that the slow rise gave people many years to move away to higher ground, but the archaeologist Luc Amkreutz of the National Museum of Antiquities in Leiden, studying the chemicals in the collagen of recovered human bones from Doggerland and certain Dutch beaches made from dredged North Sea sand, thinks the moving-away process was not gradual. The larger the fens, the richer the diversity of plants, fowl and fish. Who would voluntarily leave such bounty?

Some researchers thought the underwater landslide known as the Storegga Slide of 6100 BC might have been the definitive catastrophe that supercharged the speed of the inundation. When a large underwater chunk of what is now Norway slid, it might have caused coastal tsunamis, immediately drowning out the last of Doggerland's Mesolithic seaside settlements and their people. The archaeological evidence for such a wipeout hasn't come to light and the archaeologists conclude:

Ultimately the Storegga tsunami was neither universally catastrophic, nor was it a final flooding event for the Dogger Bank or the Dogger Littoral. The impact of the tsunami was highly contingent upon landscape dynamics, and the subsequent rise in sea level would have been temporary. Significant areas of the Dogger Littoral, if not also the Archipelago, may have survived well beyond the Storegga tsunami and into the Neolithic . . .

Yet I wonder if, as the waters rose, metamorphosing proto-England from the doorstep of a vast continent to a small island, some landscape memory of hugeness underlay the country's later drive for empire.

By 2000 BC the North Sea had the upper hand. In this drenched world peat bogs began to devour the woods, enveloping twig and stem, bud and root. Centuries later when turf cutters worked the material they exhumed long-buried branches and trunks preserved in the cold acidic peat. A valued find was a recumbent pile of bog oaks from the deep past. When dried, bog oak could be burned or worked. It is still prized for its deep black color and hardness—the equivalent of rare and precious ebony.

In the 1930s North Sea fishermen hauling in their nets were annoyed to see mixed in with the fish well-preserved pieces of trees as well as bones, flints, shells and solid pieces of old peat they called "moorlog." Over the years the more observant came to understand that below the North Sea there was land once inhabited by humans and they called this productive fishing ground Noah's Wood, surmising that the biblical flood had drowned the place. Archaeologists, impressed by the prime underwater preservation of wooden artifacts, began to look at the debris hauls. They became very alert once they identified the bones of extinct animals. In 1931, the English trawler *Colinda* fishing on Brown Bank found in its nets the usual familiar bottom rubbish of moorlog, tree branches, bones.

All the unwanted debris was usually "heaved" over the side of the boat without much consideration but this occasion was different. One large piece of peat was hit by a shovel and gave a strange noise. The master of the ship, Skipper Pilgrim E. Lockwood, decided to investigate and broke open the block. Out dropped a prehistoric antler "harpoon."

Similar bone points or possible fish spearheads were known to archaeologists from Denmark and England and were identified with the shadowy hunter-gatherer Mesolithic people who lived between the stone-working Paleolithics and the gardening Neolithics. Later, when the rare Mesolithic hunter-gatherer site of Star Carr in northern Yorkshire was excavated, similar antler points turned up as well as some eerie red deer antler "frontlets"—prepared deer skulls with the antlers still attached and "eye holes" bored in the back of the skull. The archaeologist Sir Grahame Clark made his reputation in 1949–1951 digging Star Carr in a region occupied by immigrant Danes millennia after Mesolithic times. The name comes from the Danish *star kjær*, meaning "sedge fen." Star Carr has been judged archaeologically as equal in importance to Lascaux. Clark's work with the new radio-carbon dating processes of the Star Carr artifacts brought him a Nobel Prize in chemistry and gave some definition to the Mesolithic. Several decades later new archaeological surveys expanded the Star Carr site and found other carrs around extinct Lake Flixton with even richer yields of artifacts. Star Carr was settled around 9000 BC and the objects they dug up showed dates of 8000 to 6000 BC. The antler point "harpoon" caught by the *Colinda* had a radiocarbon date of 11,740 BC plus or minus a few hundred years.

The *Colinda*'s antler point date was archaeological news: it theorized the unknown residents of Doggerland might have been Mesolithic people—little-known, little-studied and little-admired. The recognition that a missing underwater land mass could have been the homeland of the Mesolithic people was tremendously exciting but equally frustrating because it was not possible to get at any submerged settlements below the North Sea. The scientists had to be content with whatever fishermen hauled up until 2001 when, during a seminar on the Mesolithic at the University of Birmingham, someone suggested that the store of seismic data piled up by oil and gas companies in their searches for fossil fuel deposits deep under the sea might also be able to pierce the shallower waters that covered Doggerland, although it was unknown if the deep-probe oil and gas data could also divulge what lay closer to the surface. The early trials were successful, and when the seismic data was combined with years of extensive archaeological underwater mapping the results were stunning. The geoscientist Huw Edwards of Petroleum Gas Services described the shock of recognition:

> . . . we huddled round one of our computer workstations and instead of looking deep down to the seismic data for oil, we applied the latest in petroleum exploration technology to the shallow section. To our amazement, for the first time in thousands of years, the long-forgotten surface of Doggerland started to appear. Science does not get more exciting than this and it dawned on us that we were witnessing the start of a new era in marine archaeology.

Since then scientists have mapped many of Doggerland's undersea features including old coastlines, sandbars, low hills and ancient rivers. This melding of marine geophysics with archaeology put a strong emphasis on underwater exploration, for cold water preserves artifacts far better than terrestrial soils. Wondrous maps of an invisible country emerged to heat the imagination. What revelations lay below? Diving exploration was not an option as the North Sea is crisscrossed by busy shipping lanes. In May 2019 scientists working from their research vessel *Belgica* taking core samples, dredging up mud and rooting around with a mechanized claw discovered a drowned forest and possible traces of a settlement in the vicinity of Brown Bank. They later recovered a flint hand tool. They also retrieved a remarkable single core that showed the transition from fully terrestrial sediment to a fully marine environment, something never before seen in one sample, and indicating a rapid rise in sea level. Mesolithic people had had to cope with rising waters from immense ice melt, as must we of the Anthropocene.

Modern Western economic history tells the story of humans' unremitting domination of every other species, of ceaseless modification and reshaping of the natural countryside to facilitate the taking of whatever is in the natural world that will make wealth. The attitude of looking at nature solely as something to be exploited—without cooperative thanks or appeasing sacrifices—is ingrained in Western cultures. A "Resolve" by a West Texas cattlemen's meeting in 1898 makes the point:

Resolved, that none of us know, or care to know, anything about grasses, native or otherwise, outside the fact
that for the present there are lots of them, the best on
record, and we are after getting the most out of them if
it means the way to continued prosperity.

Britain was the front wave of the industrial revolution, yet
that revolution was laid over a still-rural world that had built up a
rich literature and language expressing an intense love for nature,
such classics as Reverend Gilbert White's *The Natural History and
Antiquities of Selborne*, William Cobbett's *Rural Rides*, the curate
Charles Kingsley's *Water Babies*, W. H. Hudson's *The Naturalist in
La Plata*, J. A. Baker's *The Peregrine* and more recently the books
of Robert Macfarlane, Tim Dee and James Rebanks.

Fewer of today's English move "feather-footed through
the plashy fen." The early fens covered 15,500 square miles,
but centuries of land enclosure and relentless drainage, combined with a steady move to industrial urbanization, wore
away much of the common link with rural nature lore. Today
less than 1 percent of the original fens remain. General refer-
ences to the outdoor world have become rarer, despite *The
Guardian Weekly*'s "Country Diary" essays. Today anyone can
see televised nature programs spiced up with exoticism, lost
secrets and mysteries. Belief in the Landscape Beautiful lingers on in private gardens and lawns and through images of
greensward television costume dramas utilizing the once-
private great estates now held in the National Trust, that
famed charity founded in 1895 that preserves historic places
and environmental heritage.

What, then, prompted the World Wildlife Fund (WWF),
which biannually reports loss and degradation of the natural

world, to write in its painful 2018 "Living Planet" report that the United Kingdom is one of the most nature-depleted nations in the world? (Undoubtedly the lost fens were part of the tally, although Wicken Fen, one of the National Trust's first properties, was saved—a tiny bit of the once-vast fenlands.) The report cited intensive mechanized farming and degraded farmland, noted that cars kill 10,000 badgers a year and that ancient hedges were being torn out to accommodate development.

That report upset many, including the farmer-author James Rebanks, working to restructure the old family farm in ecologically beneficial ways. Just three years earlier Robert Macfarlane's *Landmarks* had resurrected British nature vocabularies of land and water and become an international best-seller. The main ingredient of *Landmarks* is its capture of words the British once used to describe rural life and land. Fen words are scattered throughout, but a sampling from East Anglia and fenland gives us "roke" for the rising of the evening fog; "skradge" for an earthen bank atop an old bank; "fizmer" for the sound of grass moving in light wind; "didder" for the way a bog quivers when one steps on it; "poise-staff" for a jumping pole to leap over dykes. J. R. Ravensdale in his 1974 *Liable to Floods* contributes "black waters" (standing flood in contrast to a moving river's "white waters"), "stulp" (a boundary marker or as a verb in the act of marking a boundary) and "roddon" (the curious raised bed of a vanished river, raised because old silt deposits filled in the course and the shrinkage of adjacent drained peat threw the eerie ancient shape of the watercourse into sharp relief).

It is a given that the loss of natural places and discarding of their vocabularies have multiplied over the last hundred years at exponential speed, not just in England but in most countries. Every place has its own story, but the persistent and fatal draining of the English fens over three hundred years is an example of how key parts of an ancient natural countryside were gradually but deliberately engineered out of existence as the language and the knowledge of the fens dribbled away.

What Were the Fens Like?

People lived in the English fens for thousands of years through the Paleolithic, Mesolithic and Neolithic times, the Bronze and Iron ages, the Saxon years, the Roman centuries and the medieval centuries despite the annual floods. The floods came, but they went and the richness of the commonly held region was irresistible. Excavations at Star Carr indicated that the starchy rhizomes of the reeds that grew by the millions were used and may also have been a stable food source in the fens, a precious fallback if unexpected catastrophe struck or hunting, fishing or fowling failed. In the creeks, rivers and estuarine bays swam salmon, flounder, sea trout, eels, shad, sturgeon and more. No one would starve but the ills of ague and malaria were risks that came with the territory. At their best times the fens were "a source of wealth that could hardly be surpassed by any other natural environment."

In the past forty years wetland archaeologists have discovered much about the people who lived in early British wetlands where water preserved wood, stone, metal and peat bogs preserved skin and leather. Dendrochronological dating gives remarkably precise dates.

When the Romans were in Britain they settled a few places in the fens during a relatively dry period. They built a road, the Fen Causeway, connecting the fens with central England (and crossing the Neolithic Flag Fen trackway), and also did some drainage, guiding excess water into rivers that flowed through the fens to the sea. Yet long before the Romans arrived the Neolithic fenlanders had built complex trackways through the fens. They were made of planks, brush, poles and other materials. The oldest known road in Britain when it was discovered in 1970 is the mysterious Sweet Trackway, built in 3807/6 BC.

The track had been built from wood from a mature primary, secondary and deliberately managed woodland with primary woodland producing the majority of oak planks and secondary and managed woodland the straight rods and poles of ash, hazel and oak. . . . We do not know where the Sweet Track comes from, or where it goes to. It remains unclear what social, political, economic or religious objectives were achieved by it.

The Flag Fen trackway in East Anglia dates to the Bronze Age over many years dated from 1365 to 907 BC. The archaeologist Francis Pryor (familiar to fans of the long-running BBC television series *Time Team*) discovered it in 1982. This important track was made with 60,000 timbers and posts, and at some distance along the walkway it connected to an island. The people using this trackway deposited an extraordinary number of objects in the surrounding water. The word *liminal* applies to the Flag Fen trackway. It stems from the Latin *limen*, meaning a threshold or entrance. In his studies of the trackways Pryor uses "liminal" to indicate a space dedicated to a rite of passage, specifically the crossing of the threshold/boundary from life to death. The objects placed in the water—broken swords, daggers, pins, spearheads, earrings, polished white stones, pots, horse mandibles—expressed votive significance.

It is likely that trackways in fens, bogs and swamps were used in many ways: to reach areas where certain medicinal plants grew, where fish or ducks congregated, to get at bog iron, to connect to other trackways and so to other settlements. Building a trackway must have given the road gang a sense of power—reaching the unreachable and knowing the unknowable.

From the Neolithic through the fourteenth and fifteenth centuries North Sea levels were relatively low. The flat infinite-skied fens covered 3,900 square kilometers punctuated by extensive high and dry areas that supported about fifty small settlements and stock farms. People stacked peat slabs for the walls of their round houses. They understood the differences of the richer minerogenic flood plain soils and the watery moors and wastes. These fen men and women built and lived on the dry ground, using the surrounding rivers, lakes, canals and bays for food and building materials. They had intimate knowledge of the creeks, rivers, mudflats, reed beds, high ground that could support houses. They ran their small cattle on the seasonally dry grasslands, caught eel and fish, took wild bird eggs. They hunted auroch and red deer and eventually began to tame wild cattle and build up stock herds. Their houses were warmed by fires of peat turfs whose nostalgic smoke later perfumed many pages of Irish literature. The fenlands were a remarkably flexible habitat, a diverse mix of water and land that seasonally and annually changed, fostering inventiveness and trial-and-error experimentation in its inhabitants. That all changed.

From the fifteenth century onward fen history is one of smallholders, rural commoners and fenlanders who year after year were eased out or even forced off the lands traditionally held in common. The new "owners" understood very well that wealth came not from eels and bundles of reeds but from writs and deeds, from dryland wheat crops, cattle, sheep and large-scale increasingly mechanized farming. The loss of traditional multipurposed areas damaged the ebb and flow nature of living in the fens. The 1872 publication of a census, the "Return of Owners Land," showed that less than 1 percent of the population owned more than 98 percent of English land. In the twenty-first century Lincolnshire in East Anglia, once prime fenland, shows flat mono-farms of cereal grains and looks very like the giant fields of the American and Canadian midwests.

We rarely think of the sounds of ancient days but here is the Victorian writer Charles Kingsley's often quoted vivid description of market gunners in Whittlesey fen:

. . . dark green alders, and pale green reeds, stretched for miles round the broad lagoon, where Coot clanked, and the Bittern boomed, and the Sedge-bird, not content with its own sweet song, mocked the notes of all the birds around; while high overhead hung, motionless, Hawk beyond Hawk, Buzzard beyond Buzzard, Kite beyond Kite, as far as the eye could see. Far off, upon the silver mere, would rise a puff of smoke from a punt invisible from its flatness and its white paint. Then down the wind came the boom of the great stanchion-gun; and after that sound another sound, louder as it neared; a cry as of all the bells of Cambridge, and all the hounds of Cottesmore; and overhead rushed and whirled the skeins of terrified wild-fowl, screaming, piping, clacking, croaking, filling the air with the hoarse rattle of their wings, while clear above all sounded the wild whistle of the Curlew, and the trumpet note of the great wild Swan.

They are all gone now.

High above the East Anglia freshwater peatlands lived the chalk and limestone people in their towns and cities. They thought of the downslope fens as one vast spread of reeds and grasses full of unpleasant and sickly paupers tottering around on stilts or poling homemade boats, a place good for eels and a day's bird-shooting but otherwise a watery waste.

The reality was different. The fens were a diverse mix of solid ground and seasonally wet reed beds, water deep enough for sturgeon and shallow enough for eel traps. On the higher islands the residents built their houses and worked gardens. There were ancient pingos, high-banked circular water features formed by the action of permafrost or freezing lobes of water in glacial times, later transformed by melt to placid lakes. The Breckland meres were a series of ponds, depressions in the chalk whose water levels still fluctuate in peculiar ways as they are connected underground to a deep and changing water table. Claypits were used by fen inhabitants from early times to make pottery, in daub-and-wattle house construction and dam construction. The residents likely found that life in the fens was the pig's whiskers. When the intruders came with their platitudes about "betterment" and "improvement" the locals fought against the changes but were outmatched. Although they understood the fens in a way the projectors never did they lost their battles one by one over the generations.

The fen people of all periods knew the still water, infinite moods of cloud. They lived in reflections and moving reed shadows, poled through curtains of rain, gazed at the layered horizon, at curling waves that pummeled the land edge in storms. I wondered how the fens looked and consulted the works of Dutch painters who lived in a similar environment of water and sky and wavering horizon. These regional artists developed an acute sensibility for the subtleties of grey, for landmarks and evanescent levels that fade from sight or emerge from the wet air. Fen and polder beauty could be as delicate as "the gas of moonlight." But in the fens there was color as well. A visual journey into the modern peatland is possible through the photographs of Wolfgang Bartels, whose specialty is *Moorlandschaften*, the bogs and fens of northern Germany, saturated color and close-up detail as well as the characteristic melting horizons and vanishing perspectives punctuated by ditches and dikes, pathways, paddocks and berms. Tufts of swamp grass in summer take on frozen porcupine shapes in winter, the black arms of drowned forests protrude from the water. There is a flare of red willow, a sense of the black wet underworld.

The fen people were poorer than uplanders in luxury goods and fashionable social amusements. Certainly ague and hunger and malaria were known. But they lived in one of the world's richest environments, a grand estuary habitat for beaver, water voles, many kinds of ducks and geese, eels, ospreys and marsh harriers, passerine birds and the huge common crane four feet tall stalking about with a watchful eye for frogs, innumerable butterflies and moths, vast numbers of whirling iridescent dragonflies and myriad plants—grasses, willows and reeds, marsh orchids, the yellow flag iris, fen violet, the round-leaf sundew and cranberry species—plants that disappeared with the fens.

From prehistoric times to the nineteenth century the waterlands were the gardens and grocers of the local population, their butcher shops and fish stalls, their highways to trade and market centers where they could exchange or sell their fenny produce which might include the water reeds that made superior roof thatch that could last fifty years, and rushes for lighting. Because the rushes (*Juncaceae*) grew in the millions, evenings in a medieval fen dweller's home could be warm and smoky yet well-lighted. The rushes, upwards of thirty inches in height, were gathered by women and children and dried, then peeled to reveal the inner pith except for a strip of the exterior skin left in place for strength. The rushes were soaked in animal fat and further dried, then burned in special holders as the nineteenth-century English garden designer Gertrude Jekyll described:

> About an inch and a half at a time was pulled up above the jaw of the holder. A rush-light fifteen inches long would burn in about half-an-hour. The frequent shifting was the work of a child. It was a greasy job, not suited to the fingers of the mother at her needle-work. "Mend the light" or "mend the rush" was the signal for the child to put up a new length.

Fenlanders observed and worked out how to manage the wetlands, how to repair and augment natural banks after river flooding or rising sea waters or heavy rains. Over millennia of occupation they became water experts, managing flowage and breaks, understanding annual weather cycles, ditching and diking to prevent flooding or to hold water.

Some contemporaries suspected that the fenland residents enjoyed a society freer from the rigid class and wealth divisions that distinguished urban uplanders. With so many resources ranging from rich pasture grass for livestock and the resultant milk, meat and leather, legions of tasty eels and fish, wild ducks, furs and feathers and peat fuel even the poorest fen dweller could make a living and sell the surplus to inland towns accessible by stream and river. But the fenlanders were neither saints nor the oft-described cretinous lazybones; they were people who wanted their way of life to stay in its comfortable traditional ruts. They had an intense dislike of outsiders. The dislike was reciprocated.

They also suffered much illness, most often rheumatism, "the ague" and "swamp fever"—the last two sicknesses left their victims incapacitated and reeling if they did not kill them outright. Traditionally when fen people were ill they doctored themselves with the medicinal plants around them. The ever-useful rush, when steeped in water, had soporific qualities of interest to researchers today. Brookweed (*Samolus valerandi*) was good for scurvy and healing wounds, common nettle reduced inflammation of the mouth and throat, bog rosemary (*Andromeda polifolia*) lowered blood pressure (but had dangerous side effects). Botanists have identified ninety-four fen plants in the area of the Whittlesey Mere; a quarter of them are now extinct.

Upland dwellers literally and metaphorically looked down on the fen dwellers and called them ignorant, brutish and diseased. A description of the fen country residents attributed to Sir William Dugdale, the seventeenth-century authority on fen drainage, reads, "a vast and deep Fen affording little benefit to the realm other than Fish and Fowl with over much harbor to a rude and almost barbarous sort of lazy and beggarly people." A chief complainant was a once-wealthy eighteenth-century historian-topographer, Edward Hasted (1732–1812), whose four-volume *The History and Topographical Survey of the County of Kent* is today a rare curiosity. Hasted is described in the *Dictionary of National Biography* as "a little, mean-looking man . . . imprudent and eccentric" and embroiled in "pecuniary embarrassments." The biographer both slammed and praised Hasted's topographical survey saying that although it was "very defective in details of social history and in biographical or literary history . . . [it was] a faithful record of the property of the county and of the genealogies of its principal families." It was also packed with vivid descriptions of the "Unhealthy Parishes" of the county, listed in Mary J. Dobson's magisterial *Contours of death and disease in early modern England*. Here is a sample of Hasted's comments on fen parishes:

. . . a very forlorn unhealthy place . . . much subject to agues . . . the air is very unhealthy, and much subject to intermittents, a fatality which attends in general all these parishes . . . badness of the water and the gross unwholesome air . . . agues, which the inhabitants are very rarely without . . . unhealthy to an extreme, the look of which the inhabitants carry in the countenances . . . this town cannot possibly be healthy . . . fogs and noisome vapors.

It has to be the oldest story in the world—taking "worthless" lands from people deemed defective and inferior. The naturalist Ellen Meloy, in *The Last Cheater's Waltz*, commented when reviewing various countries' choices for nuclear testing—"the terrain of strategic death"—"Common to the lands is a consensus of their worthlessness and the assumption that local populations were invisible, expendable, or relocatable."

In the early medieval period the five monasteries that claimed land in the fens grew richer and more powerful. The busy monasteries often served as hostels for travelers and many fenlanders worked for the orders, paying rents to them— most commonly in eels. The Crown also held fen property, including Hatfield Chase, then a royal hunting forest. Over the centuries small local drainage projects grew into a bewildering administrative tangle of tithes, fees, rents, duties, responsibilities and usages. From the Middle Ages on as the North Sea gradually rose and the lowest of the dry areas took on water, the fen people at first found ways to divert water from new-flooded winter cow pasture. Although they were deeply knowledgeable about banking and diking, closing off and opening flowage, the rising North Sea levels made drainage increasingly difficult to manage, giving an entrée to drain-the-fens acquisition men and professional projectors and foreigners who did not understand fen waters but fully grasped the worth of dry fertile real estate.

A keen interest in increasing agricultural and grazing land came with the expanding population after the Black Death. Flooding increased in the time when Henry VIII outlawed the monasteries which included the five major landowners in the fens, and redistributed their fenlands to his cronies and followers. People with no understanding of the annual maintenance of the fenlands moved in. In 1604 the first of thousands of Parliamentary Enclosure acts to transfer common lands (including fenland) to private ownership began and by the seventeenth and eighteenth centuries land "improvement" schemes dominated land discussions.

Sometime after the Stuart King James I came to the throne in 1603 once again there were devastating floods in the east coast fens. James magnanimously vowed to undertake a massive drainage project. (The fens have always inspired "projects," first drainage, today rewatering aka "paludification.") He hired one of the foremost wetlands engineers of the period, the Dutchman Cornelius Vermuyden, experienced in Low Countries drainage. The first target was the Northern Fen, including Hatfield Chase, an area of 59,000 acres. Years passed and in a stroke of fate James died of "marsh fever" before the work started, but his heir, Charles I, proceeded and in 1626 a contract was finally signed. Vermuyden was to receive a staggering payment for his work—one third of the reclaimed land. He brought in Flemish workers for the project. When the local fensmen saw that not only were they not given jobs, but they would also lose their fen-commons rights, they rioted, tore out the Flemish work again and again, stoned the foreigners. Vermuyden finally left this work unfinished and also left behind many of his Flemish workmen. A new river—the Dutch River—had to be cut to repair the damaged fen. During the English Civil War (1642–1651) dykes were deliberately breached and sluice gates raised causing a vast flood that undid all of the drainage works of earlier generations. Alas, Vermuyden had not returned to the Low Countries. He was busy in the Southern Fens, draining the Great Level in Cambridgeshire. Again there were riots and harassment, songs and ballads. Popular was "Powte's Complaint," "All will be dry, and we must die."

Malaria was the ogre of the fens, and the English uplanders who lived in a time of awful stenches thought the fens emitted the worst of all odors. The illustrations in Skertchly's *Fenlands* shows an intriguing and beautiful lost world, but something we cannot see is the frightful stink associated with the fens.

Dobson begins her *Contours of death and disease in early modern England* with a rampantly adjectival "Olfactory Tour" of the period.

> We are confronted with places of "a thousand stinks," airs of overpowering nastiness, waters of stagnant and stinking mud, hovels of putrefying decay, cities of foul and filthy fumes, effluvia of rotten human and animal flesh, streams of sickly stenches, alleys of corruption, and noisome corners of festering filth. We are offended by the smells of stinking breaths, the descriptions of foul spittle and black vomit, the scenes of unwashed bodies crawling with nauseous and venomous vermin, the sight of human and animal excrement in every corner, the exhalations of lousy men, women and children. . . . Amongst the most foul and fatal of all airs and waters in the natural environment were the bad airs and stagnant waters of low-lying marshland districts . . .

This description rather scotches the idea of time travel to Ye Olde England for me. The marshland smell, of course, was hydrogen sulfide, corrosive, poisonous and flammable. It is the smell of the process of microbes digesting organic material. During the fen drainage work when soil and mud were stirred up it gave off this stench. Undisturbed areas of the fens were scented by water lilies, reeds and general fishy-eely-birdy smells except where gaseous bubbles burst at the surface.

Malaria: the name is Italian for "bad air"—*mal aria*. The most common malaria strain was the killer *Plasmodium falciparum*. More benign in comparison was *Plasmodium vivax*, also known as "the ague" and "Tertian Fever" which earned its name by spiking its victims' fevers every third day. (A variation on this rhythm is the malarial illness experienced by one of those English explorer types once so plentiful—Clifford W. Collinson F.R.G.S., who described in his 1926 *Life and Laughter 'Midst the Cannibals* how "after a nightmare night of burning heat and shivering cold, I suddenly broke out into a profuse and most luxurious perspiration . . . [and] throughout all the years I spent in the islands these short and sharp attacks of fever recurred regularly at intervals of about three weeks . . ."

In the eighteenth and nineteenth centuries many fen families stored premade coffins for the inevitable end, and burials in this part of England outnumbered births. Rotherham in *The Lost Fens* believes "the Roman Legionnaires brought the disease to Britain and so to the Fens." Dobson thought it endemic to England from very old times. Dobson's gathering of comments on the pallid and short-lived fenlanders makes sad reading, though many built up enough resistance to the diseases to keep alive into adulthood.

Daniel Defoe, in his 1727 *Tour Through the Whole Island of Great Britain*, told a nasty story, saying that in the fenlands he had met a farmer who was on his twenty-fifth wife. His thirty-five-year-old son had had fourteen wives. Both men, born and bred in the fenland, "did pretty well with it" [the ague] and for diversity chose their new wives from the high country. The upland women, having little or no resistance to the fenlands' ague, usually died within a year or two. This account sticks in the mind. Over the distance of two hundred years we can feel for the unknowing women ensnared by cynical fensmen who saw them as little more than sexual receptacles or stand-in caretakers for an earlier-hatched brood of motherless children.

It was not only commoners who suffered. King James I, who invited Vermuyden to drain the fens, fell ill with "marsh fever" in March 1625. His condition worsened and he "had a poultice applied to his abdomen, but this appeared to cause a series of fits, and later he began panting, raving and had an irregular pulse. Further treatment made him complain that he was burning and roasting." A rather gruesome account of his postmortem noted, "His skull, because it was so strong, was broken open with a chisel and a saw. . . . it was so full of brains . . . they could not . . . keep them from spilling, a great mark of his infinite judgement [*sic*]." Charles II was another unfortunate victim. He awoke one morning in 1685 feeling unwell. His doctors gave him the full monty: bleeding, doses of medicine, emetics, enemas, head shaved for some red-hot cauterization, noxious plasters on his feet, another laxative and more blistering agents, white hellebore to provoke sneezing and "the powdered skull of a man that had died but never been buried." The torture finally ended with his death.

Everyone, from fen men to London medicos, believed that
stagnant fen water gave off "noxious and pestilential vapors"
that caused fevers and agues. Anything that stank that much
had to cause illness. In 1878 Miller and Skertchly agreed.
They could not know that the cause was not poisonous vapors but the slender anopheles mosquito, as Dr. Ronald Ross
of the Indian Medical Service discovered in 1897. Malaria
was eventually extirpated in England but at the cost of the
very great loss of the fenlands.

Every human wanted wheat for bread—wheat and barley
and more land to grow them. These grains originated in the
dry uplands of Turkey; such cereal crops did not do well in wet
low country. But the fens were modified to cattle pasturage
and then to wheat fields through the intense drainage schemes.
The conversion of wetlands to cropland increased methane
and CO_2 output and our species increased the pace until we
are now, as someone has said, locked into a global economy
that seeks to convert CO_2 into money as fast as possible.

For centuries the entire fenland region had been a pulsing
flexible system that was never static. But in the background
lurked the uplanders who thought not of expedient and temporary drainages nor accommodating the North Sea's vaga-
ries, but massive drainage works that would convert the entire
fenland into one great region of fields. They had their way
and the fens became England's bread basket. The old rhyme
says it all:

> The law doth punish man or woman
> That steals the goose from off the common,
> But lets the greater felon loose
> That steals the common from the goose.

The three-hundred-year struggle of the British government and ruling class to drain the fens for farmland and make
the fen men into productive agricultural laborers is a classic case of misjudgment on one side and stubborn refusal
to admit defeat on the other. Again and again did fen men
riot and revolt against the drainage works, against the Dutch
experts brought in to design and oversee the project, against
this interference in their lives and traditions. Over centuries they had made their water world livable through fenland practices with osiers, reeds, duck feathers and eel skins,
but smarted under insults as upland improvers and projectors
described them as stunningly ignorant, afflicted by malaria,
given to dosing their children and themselves with opium.
They fought back but their resistances were crushed and in
the long run the sweet days before drainage when the fens
were fecund ecosystems were gone.

Most of the early wetlands were drained and plowed for
extensive agricultural fields of soils that are still rich but emit
CO_2 when tilled, can only be maintained by intermittent
flooding and have vast systems of pipes and pumps needing
constant repair. The fine silt soils tend to blow away when dry
and erode when flooded or flushed. The cost of such upkeep
and the evidence that we could reduce CO_2 emissions by
converting these artificially dry lands back to bog, fen and
swamp makes wetland reclamation a compelling interest that
inevitably butted up against a now-entrenched agricultural
hierarchy. In the effort to staunch the outflow of CO_2 and
methane gases the European Union forbade the cutting of
peat in 2011. Ireland fought the ruling until 2018 when it
gave in and raised the banner of "fighting climate change."
But in the natural world you can't easily put back what is

gone. Peat-making is a process of millennia; peat mining a matter of weeks or years.

Yet some fens may return as water levels rise and rains increase. Not all of the news of England's natural history is dark. In 1995 the Royal Society for the Protection of Birds acquired 740 acres of drained fen farmland on the Norfolk/Sussex border and set about making it a watered fen of reed beds and grazing marsh by reshaping ditches and using controllable sluices—voila! the Lakenheath Fen Reserve. Birds and wildlife have found it, and the four pairs of Eurasian reed warblers there in 1995 exploded to 355 pairs in 2002. A long list of other bird species began to arrive, including a pair of breeding common cranes for the first time in four hundred years. Other nature reserves have popped up in the drained-fen region and the common crane population in February 2021 is numbered at 200 birds.

In 2001 the Great Fen Project began in Cambridgeshire on a modest trial site of five hectares (twelve acres). The *modus operandi* is paludiculture—wet farming—to grow plants that will thrive in watery soil and provide food and dozens of useful products, somewhat as the region did in medieval times. The plants, many of them native to the fens, are bulrushes, reeds and sphagnum moss. Sphagnum moss with its fantastic ability to absorb water is overharvested in the wild, dried and sold to gardeners for improving soils. The rushes and reeds have a future use in packaging, insulation, building materials. The paludiculture project, which will rewater drained agricultural land, will hold in carbon dioxide; the benefits to birds and wildlife are the stuff of naturalists' dreams. One of the most striking parts of this project is the hundred-year time line that will span generations of workers, a breathtaking departure from the usual short-term plans of our Western minds when compared to India's Khasi people's bridge building.

In 2017 the National Trust (supported by Dame Vera Lynn, whose World War II ballad "We'll Meet Again" tinted the war years with nostalgic longing for a lost world remembered) bought 175 acres of agricultural land atop the white cliffs of Dover, restored the overfertilized soil with nutrient-absorbing barley. The chalk grassland native plants prefer low-nutrient soils. In 2019 the Trust sowed the site with a mixture of wild flowers and cereals that offered rich nectars to pollinators and seeds to birds. The new-old plants attracted bees, skylarks, buntings, peregrine falcons and butterflies, including the Adonis blue, red admiral and dingy skipper.

The restoration news was so cheering that I had to remind myself that beyond Vera Lynn's wonderful meadow and the rewatered fen reserves the reality is a world plagued by melting permafrost, sea rise, unmanageable fires that burn even rain forests, terrifying storms including tornadoes and derechos and a sharp decrease in animal and insect species. Some restoration efforts fail or will need millennia to take effect.

In the nineteenth and early twentieth centuries the United States drained most of its fenlands but a few in the most sequestered spots remain. There is one high in the Colorado Rockies of the Holy Cross Wilderness—so-called for Thomas Moran's 1875 painting of a mountain with rock fissures on one side that take the form of a gigantic cross beaming out Christian values when packed with snow. Below the mountain are spongy fenlands fed by glaciers. This high-altitude wetland is a rich habitat that supports elk, deer, countless ducks and birds, amphibians, beaver, rare plants and insects. But Colorado has been highly successful in attracting tourists, skiers and hikers and new residents to its growing cities along the Front Range, especially Colorado Springs, with its flavor of conservative government. New development needs a good water supply and there is a plan for a new dam and reservoir that will capture the Holy Cross fenlands water.

> To make a new dam and reservoir more palatable, the cities are exploring unprecedented "mitigation" of digging up and physically removing the underground fens, then hauling them and transplanting them elsewhere to restore damaged wetlands.

Yet the regional office of the Interior Department has insisted that the fens are irreplaceable and the idea of moving them is "not thought possible."

The struggles in Iraq to restore the Marsh Arabs' five-thousand-year-old wetlands drained by Saddam Hussein in 1991 are slowly working. Much of what I knew about this region came from Wilfred Thesiger's famous account of life in the southern marshes of Iraq—*The Marsh Arabs*—which appeared in 1964. (My worn secondhand copy contained a faded Polaroid photograph of the so-called thrill-seeker gentleman at the podium apparently inserted by the photographer as a bookmark.) In 2003 restoration work of the marshes began under the care of the American-Iraqi environmental engineer Azzam Alwash, who remarked, ". . . eighteen years later and I'm still at it. . . . this is not an easy project; . . . five great [oil] reserves are underneath the marshes . . ." The laborious and extremely frustrating efforts to repair Canada's maimed tar sand peatlands are not yet working. There is high hope for the Cambridgeshire paludiculture experiment but we are finding that restoration and repair of damaged wild-lands, whether fen or tropical forest, is immensely difficult: not impossible but very very very difficult to put the *genius* back in the *locus*. Humans are exceedingly good at construction and destruction but pitifully inadequate at restoring the natural world. It's just not our thing.

3.
Bogs

Let Sleeping Bogs Lie by Remco de Fouw (1990)

There is an old Chinese proverb: "If the bowl be square, the water in it will also be square." Today most of us are carelessly familiar with water—we see no mystery in it, only usefulness, ownership rights or aesthetic qualities suited to human interests. But if I let my imagination slide back to shadowed prehistoric centuries I can see that water's limpidity, its way of changing shape, its magical reflections, its apparent dark color which instantly becomes transparent when scooped up might be taken as proofs of its mystical transformative power. Water is the original shape-shifter. If I pour water from a round pitcher into a square bowl and observe that the water *has made itself square* the act can still be awesomely revelatory. I catch the edge of the idea that in early human times votive offerings accepted by revered waters may have had the force of a deeply serious contract between the giving suppliant and the receiving supernatural power. It is water that makes the world's diverse wetlands, water that holds thousands of offerings and valuable gifts. I recall Norman Maclean's final sentence in *A River Runs Through It*—"I am haunted by waters."

There are other substances that may have filled early people with wondering awe, such as obsidian, the black volcano glass that reflects like a dark mirror but when struck throws off transparent flakes of terrifying sharpness. Birds, too, with their calls that sounded like strange speech could fly up and up and up where no person could go. Were they not messengers to sky powers who understood their mysterious speech? "The language of birds is very ancient, and, like other ancient modes of speech, very elliptical: little is said, but much is meant and understood."

Peatland names and labels have become so mixed in popular use that bogs are often called moors, mires, quagmires and other names. "Moor" has several shades of meaning, particularly in Great Britain, where Vikings settled the northern upland regions. In the south the Old English word "moor" means low-lying peatland. Joosten *et al.* explain:

> The Vikings brought with them the Nordic term "mór" meaning "sandy plain," "open forest area," . . . a word sounding exactly like the Old English term for "peatland." Their identical sound and the fact that both refer to open landscapes allowed the words to merge. . . . So while "moor" remained the term for a low-lying peatland in the South, in the North the sense of "moor" shifted to upland bog and heathland sites, leaving the mires in the northern lowlands without a distinct term.

In *Kidnapped* Robert Louis Stevenson caught the essence of the upland ground in the tense account of young David Balfour and Alan Stewart fleeing across the moor ". . . broken up with bogs and hags and peaty pools . . . burnt black in a heath fire."

In Canada and some of the northern United States "muskeg" means "bog." "Muskeg" is entwined with North American history. It was an Algonkian word both for bog and for sphagnum moss: the Cree said "maskek" and the Ojibway "mashkig." Adrienne Mayor, scholar of paleo-folklore, says that the Abenaki spoke of *meskag-kwedemos*, enormous and frightful horned water monsters that lived in bogs, rivers and lakes listening for incursive humans whom they seized and ate. In Lake Superior Provincial Park one of the famous Agawa pictographs shows a war canoe paddling near a horned water monster. All who feared its slavering jaws took care to paddle in silence. Native Americans have been finding fossil bones for thousands of years; they have identified and woven them into mythologies with an evolutionary slant, telling, for example, stories of the huge mastodon and mammoth bones and teeth that were the remains of the ancestral Grandfather Bison created by the Great Spirit in the deep past, the time when giant animals tramped the world.

In the two-century-long run-up to the first global war—the "Seven Years War" (1756–1763) that entangled the Dutch, French, Spanish, British, American-born natives and settlers was called the French and Indian War in North America—both the French and British greedily explored the continent, each staking claims to the land and rivers with tree blazes and metal plates nailed onto trees, noting everything that seemed of value. Each side hired Indian allies to help them in the early battles for control of the resource cornucopia of North America. The stunningly beautiful Ohio Valley, with desirable soils, fur animals and hardwood forests, was a major prize. Although they could not know it, the rivalry would lead to the find of an important fossil in a bog that made paleontological history.

In 1739 French interests were represented by two men of the important and numerous Le Moyne family: Charles Le Moyne, the second Baron de Longueuil, was a major of military troops in Montreal, and his uncle Jean-Baptiste Le Moyne de Bienville was the colonial administrator of New Orleans. Bienville was struggling against the Chicachas (Chickasaws), proxy fighters for the British. Longueuil, with 123 French soldiers and a mixed group of several hundred Indian hunters and guides headed for New Orleans to help Bienville fight the Chicachas.

This trip was not unusual. North American Indians made immense voyages on the great connecting river-and-lake waterways as a matter of course, and a journey from Montreal to New Orleans was not a new venture for them; they knew the terrain and water routes. In their war canoes the Longueuil expedition left Montreal in July 1739 paddling west on the St. Lawrence, then across Lake Ontario, connecting with Lake Erie and then turning south into the river system which would take them to the Mississippi, down to New Orleans and the Chicachas.

The expedition camped on the Ohio River and the day's assigned hunting party of Indians headed to a natural salt lick used by animals since prehistoric times. The "lick" was a salt spring, one of the largest in eastern North America. Hydrostatic pressure forced brine up through faults in the limestone below. At the surface the wet ground around the spring was soft. So many animals came for the vital salt that the constant trampling churned up the deep sucking bog lightly disguised by a cover of opportunistic plants. Heavy animals like mastodons stepped onto what looked like solid ground. Two steps and they sank into the deep gluey mud where they lunged and struggled until they died of exhaustion, becoming part of the fossil boneyard.

On this day in 1739 the hunters found an enormous femur that gave the location its future name of Big Bone Lick, a discovery considered the beginning of North American paleontology. The hunters carried the femur, two ivory tusks and two teeth back to the main camp, where everyone marveled and Le Moyne took them into custody. The expedition continued down the Mississippi to battle; they lost to the Chicachas, who were not defeated for another ten years. Le Moyne and the fossils made it to New Orleans and eventually to Paris, where the relics went to the Cabinet du Roi. All except the ivory tusks survived the French Revolution, two world wars and remain today in the Muséum National d'Histoire Naturelle in Paris.

Fen peat forms in groundwater locations where reeds, sedges, cattails, rushes and bog beans grow in mineral soils. The plants around the edge and in the water grow, then perish, season after season, gradually filling up the fen with partially decayed vegetable matter that over thousands of years becomes fen peat. But that is not the end—there never is an end with wetlands as long as there is water. If the raised center of the tight-packed fen loses contact with the mineral-rich groundwater, it then must depend on rainwater for life. Rainwater is oligotrophic—severely lacking in minerals. This is bad news for all the old sedges and rushes relegated to the reedy lagg margins.

Yet there are plants that prefer rainwater and isolation from the minerotrophic fen habitat—they are the sphagnum mosses, heroes of the bog. Their spores are carried many miles on the wind and once the sphagnums living on rainwater

get a toehold on the ex-fen they expand upward and out-
ward in plump hummocky bulges and can form lens-shaped
raised bogs that sit atop the old fenlands peat. The process is
known by the handsome word *paludification*—and for some
time the sphagnums maintain their own private oligotrophic
water habitat. I was pleased to learn from a naturalist friend
of the 2011 discovery of Crowberry Bog, a true raised bog
in the Olympic Peninsula of Washington State. The soil that
peat makes is a *histol*, rich in partially decomposed organic
material but quick to dry out. Farmers planting in peat his-
tols enjoy bumper crops at first, but eventually the soils dry
and lose their carbon content. Deeper down in the peat layer
is the most decomposed material—muck—slimy, almost li-
quescent.

Bogs vary in shapes and types. *Raised bogs*, some concen-
tric, some excentric, are common in boreal lands. *Aapa peatlands*, also called "string bogs" and "patterned fens," are also
boreal in habitat, usually north of the raised bog area where
the fingerlike reaching of bog plants through the dark waters
looks from the air like ribbons of fraying green silk. Bogs
in coastal regions nurtured by salt spray will be showered
with varying amounts of phosphorus, sodium and chlorine.
Blanket bogs are what they sound like—thick and extensive
peat, especially notable in the United Kingdom and northwest Europe.

Scotland's Flow Country claims to be the largest still-
extant blanket bog in the world. A 4,000-square-kilometer
section may become the world's first bog National Heritage
site. The Flow Country is the major nesting habitat for most
of Europe's migratory breeding birds including 66 percent
of greenshanks, 17 percent of golden plovers, 35 percent of

dunlins, divers (known as loons in North America) and all those that depend on camouflage to raise their young without predator dangers. After millennia of evolution the colors of the birds' legs and plumage are invisible against the background of sphagnum, sedge and heathers. But in the 1980s the seemingly empty and windswept region tempted the timber-needy British into planting trees. The government granted money to drain and plow the Flow Country bogs and eventually 190,000 hectares (470,000 acres) of wetlands were mutilated into a ridge-and-furrow landscape, the ridges planted with non-native trees. It all happened so quickly that one of the conservationists trying to record the diversity of the bogs had the planters right at his heels. "We were literally running along right in front of the ploughs." They would plot and photograph all one day and the next day the machines would rip through their still-visible footprints. The tree farmers were counting their future board feet as they set out hundreds of thousands of Sitka spruce and lodgepole pine. They were disappointed when the transplants did not do well in the high-acid low-nutrient bog soils where the fierce winds common in all flat country oppressed the struggling survivors.

As the years passed the conifer plantations grew into entangling wind-bent thickets. Predators—pine martens, red foxes, hooded crows—arrived. The migratory birds, habituated to nesting safety in the Flow Country, had no experience of these enemies in this place. Angry opposition to the plantations came from conservationists and environmentalists in what has been called "one of the fiercest environmental battles in British history." By the 1990s the government subsidies evaporated in a cloud of disappointment for the would-be timbermen. A series of complex laws, EU Habitats Directives and greater knowledge of wetlands values opened a way to bog restoration. Workers cut down the languishing trees and plugged the water collector drains. The water table began to rise and bog plants of heather and bog cotton appeared on the ridges and sphagnum mosses settled into the wet furrows. This was not ideal. The University of the Highlands and Islands scientist Roxane Andersen, who was overseeing the project, did not want a corrugated landscape but a homogenous wetland. Workers began to flatten the ridges and block the ends of the furrows to hold in leveling water to wet the entire site.

After sixteen years the leveled bog has stopped emitting CO_2 and the more deadly methane and begun holding them in. But the arguments are not over and there are problems in every part of the tree-planting project from "flawed" planning to inaccurate surveys. Methane is an increasing and really scary problem. In 2020 the enormous frozen masses of methane in the Arctic Ocean began to destabilize and enter the atmosphere. Methane is eighty times more powerful than carbon dioxide in its ability to push the warming of the world. We must somehow reckon with it.

The treeless Ceide Fields in County Mayo is also a blanket bog, one that covers a Neolithic site of farm fields, megalithic tombs and the fallen trunks of pines from the ancient days when Ireland was richly forested. The Netherlands, before the massive drainage projects of the sixteenth and seventeenth centuries, was called by an anonymous pamphleteer "the great Bog of Europe, not another such Marsh in the World, a National Quagmire that they can overflow at pleasure." The "overflow at pleasure" was a reference to the Dutch practice of flooding certain lands as a defensive measure against enemies, as in 1573 when they stopped an invading Spanish army by opening dikes that submerged the strategic area.

The West Siberian Lowlands make up the Great Vasyugan Mire, with Brazil's Pantanal, the two largest remaining bog complexes in the world, 1,800,000 square kilometers, which is roughly 400,000 square miles of extremely extensive bog-marsh-swamp mix. The Vasyugan stretches 550 kilometers east to west and from north to south about 270 kilometers and contains about 2 percent of earth's peatlands. The eastern half contains the Ob and Irtysh rivers. It is edged with forests and has been suggested for World Heritage nomination although the western part has been industrialized. The peatlands of Canada and Europe are also part of the important high-latitude carbon sink. In the tropics the largest peatlands are the damaged peat forests of Asian Indonesia, and Western Amazonia. Recently discovered (2012), the intact Cuvette Centrale in the Central Congo Basin is the largest tropic peatland at 56,000 square miles. Already poachers and miners are eyeing it hungrily.

Just as the fens of England were drained and converted to crop-growing, in North America the midwestern corn and wheat belts and parts of the Central Valley of California were once peatlands that were drained, plowed and planted. They too have been releasing methane and carbon dioxide since the drainage ended and although they look like lush croplands (thanks to commercial fertilizers) they are still billowing out the invisible gases. Some small undisturbed fens and bogs do survive in the rural United States, notably across New York, Vermont, New Hampshire and Maine. I once lived on the edge of the boreal lowlands of Vermont's many-bogged

Northeast Kingdom. Yellow Bogs is riddled with interlocking boglets and black spruce swamps in the Nulhegan basin so prized by dragonfly enthusiasts. At the time of my visit to the bogs I was not awake to the allure of sphagnum mosses and only years later realized the bounty of species that grow in the Yellow Bogs: *Sphagnum fimbriatum, S. wulffianum, S. rubellum, S. fuscum, S. magellanicum, S. recurvus, S. flexuosum, S. quinquefarium* and *S. angustifolium. S. fuscum* is a classic marker for low-nutrient habitat.

The word *bog* comes from the Gaelic *bogach* and the earliest literary use seems that of the Scots poet William Dunbar in his 1505 work "Of James Dog." The man he wrote about, James Dog, or Doig, was the keeper of the Tudor Queen Margaret's "wardrobe," the word not only for clothing but a wide range of royal belongings. Dunbar's teasing poem was considered witty with its play on the wardrobe keeper's name.

> Quhen that I speik till him freindlyk,
> He barkis lyk an middling tyk,
> War chassand cattell throu a bog
> Madam, ye heff a dangerous dog.

Suspense writers find bogs very useful. Bogs stir fear. They are powerfully different from every other landscape and when we first enter one we experience an inchoate feeling of standing in a weird transition zone that separates the living from the rotting. Black pools of still water in the undulating sphagnum moss can seem to be sinkholes into the underworld.

The scientific description "Palustrine Emergent Wetland" can never fit the fictional and fatal Grimpen Mire where "a false step . . . means death to man or beast" in Conan Doyle's "The Hound of the Baskervilles," a story inspired by the author's visit to the notorious Fox Tor Mire. In the story Dr. Watson hears from the naturalist Mr. Stapleton that the green-blotched apparently grassy highland is not what it seems. It is an upland moor dotted with bogs that swallow up the incautious free-roaming ponies which are sucked down with "a dreadful cry." H. H. Munro (aka Saki) used an upland moor setting to destroy the mental stability of the neurasthenic Framton Nuttel in one of his best-known stories, "The Open Window." Vladimir Nabokov's story of lepidopterists in a bog—"Terra Incognita"—also strikes the sinister note.

> Around it grew golden marsh reeds, like a million bared swords gleaming in the sun. Here and there flashed elongated pools, and over them hung dark swarms of midges. . . . Gregson swung his net—and sank to his hips in the brocaded ooze as a gigantic swallowtail, with a flap of its satin wing, sailed away from him over the reeds, toward the shimmer of pale emanations where the indistinct folds of a curtain seemed to hang. *I must not*, I said to myself, *I must not.*

Delirium descends on the lepidopterist before a moment of reality when he understands that it is urgent to get away from this wetland even though "rare, still undescribed plants and animals . . . would never be named by us."

. . . And there is a hoary old mire joke: A young man was walking carefully near a mire when he noticed a lustrous top hat out near the center. Not one to pass up an expensive hat free for the picking up he gingerly made his way to it. He lifted the hat and was shocked to see the unhappy face of a mustachioed gentleman up to his chin in mud. He started to pull the man out but could hardly move him an inch. "Wait a moment," said the man, "until I get my feet out of the stirrups."

The primordial intensity of the bog's unmoving tannin-dark water and massed sphagnum seems the true element, the stuff from which ancient universal humanity arose and to which it must return. It is unnerving on a hot summer noon, the sun reverberating off the glinting water, reeds and grasses on the edge stiff as though welded and every cubic inch of air occupied by a hungry winged insect. In early morning vast mists rise from the cold night surface perverting the landscape with optical illusion that fades out color and landmark.

Painters, sculptors, photographers, poets, archaeologists, storytellers, ecologists, botanists succumb to the allure of the bog world, where moss makes its own ecological habitat, trees dare not put down roots, predatory sundews and pitcher plants eat living swamp meat, where bog cotton "breathes" through its air-channeled stems. Everything seems to lurch slightly, to sink and rise fractions of an inch. Decomposing plant material underwater sends up stinking gas that produces the mysterious lights that wobble through evening mists—the famous will-o'-the-wisp or *ignis fatuus* (fool's light). In sunlight there is the swamp sparrow's rapid iteration like a gear in your brain spinning loose. This profoundly unfamiliar setting is not so much a place as the sudden shock of perception of threatened existence, a realization streaked with anxiety. And yet there are people like the moss expert Robin Wall Kimmerer who prance happily across the squelching brocade at the risk of losing a boot. The historian of place John Stilgoe was not such a one—he felt its wild darkness:

> A fen, bog or loch is not easily known. Although travelers may easily scan the surface of such open wilderness, the depths conceal terrors as frightful as the most loathsome creature of any imagination, the squat squalid thing that personifies the beast still snuffling about in human unconsciousness . . . Watery wilderness shelters such creatures, and like all other wilderness, forever threatens to overwhelm the land ordered and shaped by man. Wilderness is the spatial correlative of unreason, or madness, of the unhuman anarchy that informs so many folktales emphasizing the ephemeral stability of Christianity, society, and agriculture.

Today as the climate crisis begins to bite and the swelling numbers of the most populous mammal on the planet—7.8 billion people—continues to grow some recognize that it is our ever-expanding human works and vast mechanized agriculture that have flattened the wilderness and introduced us to ever-new micro-organisms, while in the last fifty years more than half of the bird, mammal and amphibian populations have dwindled into memory or teeter on the edge of the extinction cliff. It is our species that seems deranged in its blind despoliations of the natural world. It is we humans who disturb millennia of secluded species with our religions, societies and agricultures, we who bring animals and their viruses out of their remote habitats and into our markets and kitchens. These days Stilgoe's beast that snuffles about in the human unconsciousness is among us—by invitation.

The student of wetlands quickly passes from wading in shallow water to the depths of complex nomenclature and liquid meanings, multisyllabic words such as "ombrotrophic" (Greek root for rain shower), "oligotrophic" (deficient in nourishment), "string-flark fens," "frost-crack mires." A magician's box of languages trade off-the-mark translations with each other. The editors of *Mires and Peatlands of Europe* mention the problem of nonmeshing definitions while vetting the studies from the many-languaged 134 contributors that make up the compendium. A changing climate dictated that they include such emerging wetlands as "recently drained lakes, devegetated wet areas, and wet grounds newly exposed by glaciers." As the glaciers and ice melt, as oceans and groundwater rise there will come a world of new estuaries, rivers, lakes, fens and, eventually—vast bogs and swamps.

It is the business of wetlands scientists to identify and catalogue the ever-shifting ways in which water, land and plants combine. Ongoing reinterpretations of wetland sites include the transitional area from dry to wet which expresses a boundary. Time shows us the earth as a fluid patchwork constantly in flux so slow it is invisible. Centuries and millennia are the hours and days of a bog.

Albrecht Dürer, the great painter of the northern Renaissance, was deeply interested in the natural world and landscape. His exquisite 1503 watercolor *The Large Turf* has delighted viewers for the past five hundred years. He painted his less exquisite but informative 1497 watercolor *Der Weiher* ("The Small Pond") after his first visit to Italy, where he found the atmosphere kinder to artists than in Nuremberg and wrote to his lifelong friend the humanist Willibald Pirckheimer, "Here I am a gentleman . . . at home I am a bum." This watercolor sketch is the first known artist's representation of a natural wetland. Dürer caught the transition stage of fen to bog showing the "wet marginal 'lagg' zone and the elevated open raised dome as distinct components of one and the same bog." The "lagg" zone is the remnant reedy fen area between the bog and the outer mineral soil—where the golden reeds grow in Nabokov's story.

Sphagnum Moss

Wet, acidic, oligotrophic nutrient-poor waters are home sweet home for sphagnum mosses. They grow mostly in the northern hemisphere. The northernmost are at 81° N, near Svalbard. Among the early bryologists who commented on the Great Dismal Swamp straddling North Carolina and Virginia was the first notable bog authority, the Swiss-born Leo Lesquereux (1806–1889), who came to North America at the request of the celebrity scientist Louis Agassiz, who had known him in Neuchâtel, where Lesquereux worked on mosses and bog plants.

Lesquereux's life was a long string of painful events, beginning with a boyhood fall from the heights of a cliff where he had been plant-hunting. His family found him at the bottom, apparently dead—but not quite. After weeks in a coma he revived and started his botanical researches again. His most intense interest was peat bogs. He was the first to understand the process of peat formation after he invented a kind of peat auger to probe the boggy sphagnum depths. His opinions were not accepted at the University of Neuchâtel until, standing on an actual bog with Louis Agassiz and the doubting scientists, he demonstrated his proof. When the government put up a reward of 20 gold ducats for the best essay on peat bogs, Lesquereux won handily with his "*Recherches sur les Tourbières du Jura*" which established his reputation as a bryologist and which became the most authoritative work on peat in his lifetime. William Darrah, writing in 1934, referred to his "phenomenal reputation as a Paleobotanist."

Whatever happiness and surety he enjoyed took a violent slam when during an illness that was affecting his hearing and sight he went to Paris for treatment. His friend J. P. Lesley, who wrote his obituary, said that in Paris he was treated by

> a noted oculist and aorist . . . with the brutal recklessness customary at that time in the medical profession of that metropolis. . . . His Eustachian tubes were burst, and an inflammation of the brain superinduced which threatened to destroy his sight. When he returned home he became stone deaf . . . to the day of his death.

He compensated by becoming a brilliant lip reader in German, French and English, and kept up his peat bog studies with a special interest in the "dear mosses." Agassiz, then working in Ohio with the wealthy bryologist W. S. Sullivant, promised him "scientific employment," encouraging Lesquereux to join the stream of scientists who fled Switzerland after the 1848 revolt against monarchies. But once in Ohio Agassiz paid Lesquereux little or nothing for his work classifying mosses and botanical discoveries. Lesquereux, his wife and sons were forced to fall back on the family's traditional calling of watchmaking. When Lesquereux visited the Great Dismal Swamp in 1853 Lesley wrote that he "compared Lake Drummond with the raised bogs of Europe . . . Years later, the theory received support from other observers." Lake Drummond is a shallow water body on a barely discernible slope—a perched hillside bog.

Lesquereux died in 1889. Darrah wrote that the fates took one more slap at him—his unparalleled and priceless collection of specimens to match his three-volume *Coal Flora of Pennsylvania* were stored "at the National Museum"—one assumes Darrah meant the Smithsonian which was founded in 1846. Darrah writes that this collection "was supposedly lost, strayed, stolen, or sent somehow to Europe."

The bryologist Robin Wall Kimmerer declared, "I know of no plant, large or small, which has the ability to engineer the physical environment more thoroughly than *Sphagnum* through the remarkable properties of the plant itself." The sphagnums are the keystone species in peatland ecosystems which hold one third of the earth's organic carbon. These profoundly managerial mosses once covered great parts of the earth slowly multiplying and dying, multiplying and dying, building ever deeper layers, encasing rocks and tree trunks, bird nests and animal bones, sucking up CO_2. But their control of a bog goes beyond sheer proliferation of plants. Sphagnums have two kinds of cells—ordinary cells with chlorophyll that photosynthesize, and barrel-shaped retort cells whose pores absorb water. When drought hits the homeland bog these specialized hyaline storage cells can release water and keep the bog moist and alive—for a while.

Sphagnum would like to conquer the world, but its low height puts it at a disadvantage. The pollens and spores of taller plants have better access to wind currents, yet the non-vascular mosses have adapted. Close to the ground the air is still—the laminar boundary. About 10 centimeters above the sphagnum the turbulent air rolls. The sphagnum is aware. It must get its spores into that transport zone. So, when the sun heats its spherical spore capsules they tighten their shape from sphere to cylinder. Cooking in solar heat, internal pressure builds up inside the constricting capsules until they explode, hurling the spores out in a mushroom cloud of vortex rings that exceed the heights of mere ballistics and put the spores in the passing lane. Most travel only a few meters, but some may catch a transoceanic long-distance ride to new territory.

Ordinary dryland plants (grasses, shrubs and trees, for example) sprout, grow and absorb CO_2. When they die and rot the CO_2 is released back into the atmosphere. But in bogs the sphagnum below the surface does not collapse and decay. As long as it is left alone it holds CO_2 and methane prisoners. Bogs and fens that were drained a hundred years ago and then plowed and planted with crops still continue to release methane and CO_2. The same release of these invisible but noxious gases accompanies thawing permafrost.

When I look at a moss-covered bog I see an undulating expanse—Nabokov's brocade; thousands of sphagnum moss heads, perhaps fifteen or twenty species, crowded together in a floating quilt. For birds and amphibians and others that live on an insect diet such a bog is a fantastic food source while caribou, moose, musk ox and other mammals take in sphagnum for its water content. A single plant may live for hundreds of years but we do not know how the sphagnums will fare as the temperatures of a heating-up earth rise. Sphagnums are sensitive to warm temperatures.

The mop-top head of the sphagnum is the capitulum, the living part of the plant. Below the surface, each stem (resembling twisted wire) extends down and down. The stem has two kinds of branches—those that spread out for support and those that dangle down to draw up water. The leaves have differing cells: the living green chlorophyllous cells which are interspersed between the larger dead hyaline cells convert sunlight into energy, and the dead cells which are the sphagnum's operative agent each hold twenty times its weight in water. The underwater dead zone of the plant can exist for years, quietly accumulating mass, becoming more densely packed and compressed by the weight of the living plant above, the plants around it and the weight of the water. As the moss top layer expands its domain, below the surface its elderly parts senesce, packing in deeper, denser, century after century as it changes into peat. It can take ten thousand years for a bog to convert to peat but in only a few weeks a human on a peat cutter machine can strip a large area down to the primordial gravel.

Within the moss there is a teeming interactive zoo where thousands of microscopic creatures live. Some are familiar from biology class—rotifers, flagellates, amoeba and the endearing "water bear" tardigrades. There are also cyanobacteria, a large variety of worms and the delightfully named sun animals—heliozoans. The Irish Peatland Conservation Council reported that a single sphagnum plant yielded up more than 30,000 tiny animalcules. These microscopic beings go into the game bags of dragonflies, bloodworms, water beetles and boatmen, frogs and caddis flies, even mosquitoes and midges. A sphagnum specialist counted as many as 50,000 plants in a square meter of hummocky bog. That adds up to 1,500,000,000 thriving bits of life in every square meter. If you are an earth-moving machine operator planning to drain bog land, think on this and resist.

Shrubs at the edge of a bog extend their roots under the sphagnum mat, supporting it for a time, but eventually they die as the moss seizes and drags them down into the acidic chill. Sphagnum releases acid and the pH of water at the edge of a bog may be as sharp as household vinegar. In the twentieth century a growing interest in the human bodies that local turf cutters had been finding for centuries in peatlands led to scientific examination of the preservative powers of raised bogs. It was believed that the lack of oxygen, the acid pH and a certain substance in the moss called sphagnol, which had antimicrobial properties, preserved the humans in the peat for millennia. But in recent years the polysaccharide sphagnan has been identified as the critically important preservative agent. It also leaches the calcium from bones, eventually softening and destroying them.

Travelers in wet, bog-rich northern Europe and North America know the mosaic of moss, heathers, cotton grass, sundew and bladderwort. John and Bryony Coles note, ". . . it is said that a footprint impressed in a moss may still be recognizable a year or more later." I thought of the description in Felix-Antoine Savard's 1943 *L'Abatis*. In the Great Depression of the 1930s Canadians suffered greatly. There were several efforts to shift poor urban families to rural places with low populations where they could grow gardens and become self-sufficient. In 1934 Savard, who was then a priest (later also a novelist and poet), led a group of ten men who were heads of families north by canoe to Abitibi to start a new parish. His hard-to-find book *L'Abatis*—the title means an area where a forest has been cut down to make farmland—is the story of that journey and its hardships. One hot noon the group stopped on shore for a midday break. Half asleep and drowsily regarding a large green fly Père Savard suddenly heard one of the men call out from the forest that he had found "*un mort!*" Examining the remains Savard guessed that during the last winter this young trapper with the red beard was exhausted and lay facedown to sleep. His arms were extended and his face against the earth in a hollow. "*Le visage a laissé son empreinte.*" Although Savard did not specify sphagnum moss as the medium that held the imprint of the man's face, it is likely, as those islands and lake shores were more bog and moss than grass.

In the northern hemisphere the use of sphagnum moss for an aseptic and soothing bandage has been known for ages. It was used for diapers by North American indigenous west coast people, used as bandage material in World War I frontline first-aid stations, is still used to keep fish and root crops fresh in Scandinavian countries. It appears even in Flann O'Brien's 1939 classic *At Swim-Two-Birds* when Sweeny, the madman poet reciting verses from his perch in a yew tree, falls ". . . a wailing black meteor hurtling through green clouds, a human prickles." Those below examine his grievous wound and the attendant Pooka gives advice. "There is only one remedy for a bleeding hole in a man's side—moss. Pack him with moss the [*sic*] way he will not bleed to death."

Scores of plants, birds, animals and insects that inhabit bogs have "bog" tacked onto their names, as bog moss, bog berry, bog blitter (the bittern), bog-lander, Bog Latin and bog butter. Bog butter is prehistoric cow's butter that was packed into a firkin, bucket or keg, then set in the bog, either for preservation or possibly as part of a sacrifice. As the stuff aged it became a pale, waxy substance with the scientific label "butyrellite." Fearfully old deposits dating back two thousand years have been pulled out of Irish peat bogs. Bog butter has been described as salt-free, pungent and no doubt was an acquired taste. Or not. And those tubs of glistening fat may have been votive offerings to gods we no longer recognize. In addition to old butter, Wijnand van der Sanden lists these objects found in the bogs: ". . . axes of flint and bronze, bronze swords and shields, iron coats of mail, vessels of earthenware, bronze and silver, bronze and gold ornaments, bronze musical instruments, wooden wagons and parts of wagons, wooden agricultural implements, wooden butter-churns, coins, clothing, balls of wool, wooden boats, plaits of human hair, wooden anthropomorphic figures and animals or parts of animals."

The wooden anthropomorphic figure van der Sanden was thinking of is likely the unique 12,500-year-old Shigir Idol, the oldest wooden mobiliary (portable) art yet discovered, made by hunter-gatherers in the late Younger Dryas period— those long aching years of drought and cold during the shift from glacial to early Mesolithic times when meltwater poured into the North Atlantic. The pioneer forests of willow, birch and pine sprang up, later replaced by elm, oak and alder as the atmosphere warmed. The Mesolithic people who shivered and starved through the cold centuries were—until recently—low on the totem pole of archaeological interest despite their persistence in staying alive. They were considered inferior to those who came later, the progressive agriculturists of the Neolithic, believed to be the innovators of civilization. With the discovery of Doggerland there was a swell of interest in Mesolithic archaeological sites and their people. Today that interest has increased: we have something in common with the Mesolithic people. They lived at a time of major climate change of the kind we are just beginning to experience. The Shigir Idol further opens up the possibilities of finding new evidence of Mesolithic people in the trans-Ural and farther east regions, long ignored but now hinting that these places may be stuffed with possible Mesolithic sites. The ancient monument also draws the narrow Paleolithic focus away from Europe and toward the trans-Ural and Eurasia. It is changing the way we see prehistoric times and people—people once presumed not to have made art did make art. The University of Barcelona paleoanthropologist João Zilhão reportedly comprehends that "absence of evidence is not evidence of absence." For me this conjures up some of the stonework of the late landscape artist Robert Smithson, who photographed a rock *in situ*, removed the rock and photographed the earthy cavity. He called the cavity "absent presence"—like Savard's look at the imprint of a man's face in the moss, like the Shigir Idol's place in art and human history.

The Shigir Idol, about seventeen feet tall when adding in the missing pieces (destroyed), was found in 1890 in the Shigir peat bog near Kirovgrad by some men hired by Count Alexei Stenbok-Fermor to dig the bog for gold. (This bog became famous for yielding up prehistoric objects and carvings.) The men brought the engraved pieces of wood to the count, who gave them to a local museum. Both the count and the statue's original lower sections disappeared during the Revolution (1917–1923).

The idol was carved from a larch tree whose rings showed 159 years of growth. It was split in half and worked with sharp stone tools. The monumental figure was incised with abstract symbols, oblique and zigzag designs somewhat similar to patterns found at the Neolithic temple site Göbekli Tepe in Turkey. Nine heads are carved on the Shigir monument, the topmost one a rather menacing face with open mouth. Dr. Thomas Terberger, the head of research at the Department of Cultural Heritage of Lower Saxony, who has studied the idol, remarked, "Whether it screams or shouts or sings, it projects authority, possibly malevolent authority." Researchers are wondering if, in addition to some possible time-stretched connection to Göbekli Tepe there might be a link to the totem poles of the Canadian and American Pacific Northwest coast that were made millennia later. I feel that many many alluring questions are yet unanswered.

Bog Bodies

In the northern European bogs peat cutters and wetland archaeologists have found not only thousands of wooden trackways, but nearby have discovered coins, jewelry, pots, tools and well-preserved human bodies. Most of the bodies were alive during the Bronze and Iron Ages. They are diverse: men, women, children, lame, sound, commoners and kings. They are the mysterious and famous so-called bog bodies.

For centuries people have pondered why those ancient humans were killed and buried in bogs. Most met their deaths through violence and more than a few were killed over and over by various means—poison plus strangulation plus drowning plus hanging—deliberate overkill that may have had a ritual significance. Wijnand van der Sanden and others believe that "many of the isolated bog bodies are to be interpreted as human sacrifices. An important argument . . . is that . . . bogs and other watery environments were evidently places where people sought contact with the supernatural world and where they deposited simple or valuable objects to put a seal on those contacts." Even in modern times bogs seem the place to deposit some bodies—American criminal gangs favored the wetlands of New Jersey for dumping corpses, including that of Jimmy Hoffa, the mob-defiant union boss. Neil Jordan's painful film *The Crying Game* was adapted from Frank O'Connor's famous 1931 short story "Guests of the Nation." In the written story the Irish protagonists are ordered to shoot two English hostages with whom they have been sharing food and playing cards and becoming "chums," because the English have shot four of their men. They are told to "collect a few tools from the shed and dig a hole by the far end of the bog." After an excruciatingly fatalistic scene the men are shot and the story ends with O'Connor's powerful sentence supposedly inspired by Gogol's "The Overcoat"—"And anything that ever happened me after I never felt the same about again." Bog bodies had and still have a ritual significance.

The researcher Bridget Brennan lists ten characteristics that describe the prehistoric bog bodies when considered as a group: excessive violence; asphyxiation by rope or garrote; ingested hallucinogens; physical deformities or other physical defects; uncalloused hands, manicured fingernails; naked or semi-naked; capes, belts, hats, armlets of leather, wool or fur; focus on head or hair; restraint stakes, withies, hurdles or bindings. But they are most often *not* considered as a group but as individual cases who might have been sacrifices or offerings to deities. Shamanism is sometimes mentioned as a likely part of the ceremonial events. The shaman is the go-between linking the everyday world with the ineffable and supernatural. The root word *saman*, from the Siberian Tungus, means "ecstatic one." The shaman could activate contact with the other world through dance, sleep deprivation, singing, hypoxia, drumming, starvation or hallucinogenic substances. The Romans were not keen on human sacrifice and when they conquered Britain around AD 60 they outlawed the practice.

The qualities of a bog that preserved human bodies, tanning the skin, halting the decay process, also saved hair, beard stubble and intact and lustrous fingernails. Even fingerprint whorls lasted after thousands of bog-submerged years and several of the bog bodies have been fingerprinted. In the sense of preservative qualities, Karin Sanders, an authority on Scandinavian literature, in her book *Bodies in the Bog and the Archaeological Imagination*, presents the bog as "a kind of ur-camera or as a pre-photographic natural darkroom."

Bodies were deposited in both fens and bogs. Van der Sanden points out that in fen bodies the soft tissues decompose but the skeleton persists. In bogs the soft tissues are preserved but sphagnan dissolves the bones. So most bog bodies become dark brown bags of skin after several thousand years.

Years ago, unable to find an English translation, I bought a copy of *Les hommes des tourbières*, a French translation of the famous 1969 *Mosefolket* by the Danish archaeologist P. V. Glob (1911–1985). There were the well-known illustrations of the handsome-faced Tollund Man, his perfect hands and feet, Grauballe Man's pressure-squeezed head, the stylish cut-leather shoes, the intricate Swabian knot of hair on a bog-stained skull.

But long before I encountered Glob's book I had been sharply interested in ancient people. I learned to read from recognizing the skeleton letters of words as my mother read me bedtime stories. It fastened my life to books and long years of endless reading. When I was in second grade I was excited to discover that the school had a library and every chance I got I rushed there to read and read until I was dragged or pushed out to the hateful recess playground. One day I discovered a startling book, the tan cover showing a rocky bluff and a cave opening. First published in 1904 it was *The Early Cave Men* by Katharine Dopp, one of America's early educators. I looked long and hard at the sophisticated illustrations by Howard V. Brown, later famous for his early sci-fi covers. I could not get enough of a drawing of two barefoot women clad in ankle-length skin dresses and fighting a bear at close quarters. One slashed with a stone dagger, the other stabbed the bear with a spear. Their expressions were intensely fierce. You can't imagine what that picture meant to an eight-year-old girl who had already noticed that in books women were always pictured holding babies, crouching over a fire or handing food to someone. Fighting a bear! The book was wonderful too because it featured a map of the cave people's country. It was the first map I had seen and it literally shaped the story. The impression of Paleolithic life that book made on me has lasted a lifetime as I observed how the general population absorbed pronouncements from archaeologists, historians and artists that emphasized the Eurocentric vision of male-dominated progressive technology. Thinking of the women and the bear I knew the questions were not all answered. Years later Glob's book also hit me hard in showing the puzzling complexity and unplumbed depths in the lives of long-gone people—women as well as men.

Glob believed that most bog bodies were human sacrifices, votives donated to the Germanic and Celtic gods and goddesses of increase; he liked Nerthus—Mother Earth. When we speak of "Mother Earth" we are still unconsciously referencing Nerthus. Glob's book opened up a strange and compelling past for the many readers who experienced a thread of connection to the ancient men and women in the peat. Artists and writers responded with especial passion to the bog corpses. Although the stories of W. Somerset Maugham set in the 1920s and '30s are less read today they flash-freeze the more unsavory English types abroad in the days of declining empire. Maugham was interested in character—aka "human nature"—and his stories often contain ruminative truisms, as in "Honolulu" when he wrote a few sentences that might apply to artists and poets on first looking into Glob's *Mosefolket*:

> It is very curious to observe the differences of emotional response that you find in different . . . people. Some can go through terrific battles, the fear of imminent death and unimaginable horrors . . . and preserve their soul unscathed, while with others the trembling of the moon on a solitary sea or . . . the song of a bird in a thicket will cause a convulsion great enough to transform their entire . . . being.

For some people influenced by Glob's *Mosefolket* the connection with the bog people became important to a life's work, as the Nobel laureate Seamus Heaney, especially his collection *North*. Novels were constructed around bog corpses: Michel Tournier's *Le roi des aulnes*; Anne Michaels's *Fugitive Pieces*; Wallace Stegner's *The Spectator Bird*; Ebba Klovedal Reich's *Fæ og Frænde*. The archaeologist Glob himself was deeply interested in art and occasionally collaborated with artists. Artists whose work plunged into primordial mud, bringing the ancient bodies out of the peat and into modern consciousness, exploring the wet underground layers, butting up against the unknowable again and again gives us a way to connect with the ancient finds. Remco de Fouw's *Let Sleeping Bogs Lie* reveals a crushed and slanted face emerging from the moss like an adult birth, like the centerpiece in a nosegay, like the nightmare that is more real than one's everyday life.

An important part of the history of the disinterred bog bodies is how they fare in our modern world. Karin Sanders does not hesitate to discuss the ethics of museum display and the dangers of the ancient people's images falling into the hands of marketers where, as Verlyn Klinkenborg said, they become trinkets and curios. In fact, the face of Tollund Man now appears on plastic pencil sharpeners and shopping bags. If we go to museums to stare at the remains of disinterred people are we not complicit in reducing the found people to curios?

I was not surprised to learn that Joseph Beuys (1921–1986) was one of the first artists to hear the call of the bogs and their bodies. In 1952 he was also one of the first to see Grauballe Man. His 1971 performance art work *Eine Aktion im Moor* was his literal plunge into a bog. The artist collective COBRA of the low countries (COpenhagen, BRussels, Amsterdam) was and is intensely interested in the bog bodies and many of their strongest works are bog related. Karin Sanders suggests that the ideas of the French philosopher Gaston Bachelard fit the ethos of the COBRA artists, especially in views of time and its passing that "the slow, notorious intimacy of the passage from liquid matter to thickening matter to matter which, solidified, bears the whole of its past within." (I overlay that thought on the mammoth gulping and gurgling in the cruel yielding mud that ultimately resulted in a huge femur on display in a French museum.) And I was led to the sense of "pastness" explored by Peter Davidson in *The Idea of North*, where it arrows in on Reinhard Behrens's invention of his mythical icebound world Naboland and the frigid waters that surround it. As the glaciers and ice melt away this sense of "pastness" may become a mnemonic refuge for humans in our hot future.

The wetland archaeologists have been busy in the bogs since Glob's book brought us literally face-to-face with people who lived millennia ago thanks to the discovery that water and peat bogs preserve artifacts far better than dry land burials. Wetland archaeology was suddenly highly rewarding as the scientists found thousands of long-submerged items in superb condition—a horse jaw, beach stones, carved wood, fragments of broken pottery, fibulae. They also retrieved preserved but mutilated bodies from ancient wetlands into the modern world where they acquired a macabre notoriety. For me, thinking and writing about the disinterred people feels slightly uncomfortable, an intrusion of their privacy. I wondered if the people who built their houses atop the coastal middens and burial grounds of Florida's ancient Muspa and Calusa people were subject to uneasy slumber. Yet I am not disturbed by the scientific examination of Egyptian and Peruvian mummies, the plaster casts of Vesuvius's victims, Ötzi, the Peruvian prehistoric people, the Makah buried under the Ozette landslide, nor the extraordinarily moving Fayoum portraits taken from Greco-Egyptian caskets. A two-thousand-year-old lump of ancient birch tar used as chewing gum with the imprint of a child's teeth in it gave me a smart sting of immediacy. At the same time that I want to know, I shudder internally at my own shameless snoopery.

One reason I am so interested in these old people is be-cause they connected with the natural world through rivers, streams, standing water, mountains, the inner recesses of caves and islands in a way we today cannot. I imagine a torch-lit ceremony in the dark hour before morning, a person with a rope around the neck led along a bog walkway to a sa-cred platform as the horizon begins to smolder red on the earth's turn and the cacophony of the dawn chorus begins. Torches are plunged hissing into the water and something important is done at the moment the day opens in splinters of sunlight. The offered person is lowered into the bog and staked down or covered with a wicker hurdle. Perhaps some people are overcome with tearful emotion, others suffused with the sweet joy of spiritual contact. In truth not I nor any-one knows the how and why of these people's temptations or struggles, their crises of belief. What we can only be sure of is that the bog under the arch of sky was its own world and that each bog body came there in the company of others yet did not return.

I was once in Banff at the same time as the traveling exhibit *The Mysterious Bog People*. Central to the exhibit was Richard Neave's 1994 facial reconstruction of Yde Girl. Every one of the excavated objects displayed interested me, but the back-ground music of an ancient *lur*, the great bronze horn made by prehistoric people, caused a shiver down the spine; the sullen bray sets up vibrations in one's bone marrow and we can guess its mournful roars reverberated through the bogs.

Thousands of preserved bodies have turned up especially in the raised and blanket bogs of northwest Europe, the oldest dating back eight thousand years to the Mesolithic. The old way of hand-cutting turfs was slow and finding a body while doing the work was upsetting. The archaeologist van der Sanden wrote:

"The peat cutter who chanced on Yde Girl with her long red hair believed he had come face to face with the devil himself. Utterly dismayed he fled from the site and didn't return to work until the next day." Had a *lur* sounded then the poor man might have fallen insensible beside Yde Girl.

Van der Sanden tells us of one of the most curious collectors of bog body information, the German archaeologist Alfred Dieck (1906–1989), who gathered material on body finds in northern European bogs from 1939 until 1986. Dieck, badly wounded in WWII and taken hostage by the Americans, continued to compile statistics for the rest of his life. He surveyed museum collections, went through files and newspapers, collected samples of hair and clothing and even interviewed peat cutters. His final count was 1,850 bog bodies. He wrote more than a hundred articles and created a vast catalogue and collection of information on bodies disinterred from bogs but, says van der Sanden,

> Everything he heard and read he copied down verbatim . . . Dieck never examined any bog bodies himself. He was mainly interested in the final reports, the paper bodies . . . It is now generally agreed that Dieck did not critically assess his sources. Everything he read or heard he took for true.

His work is regarded today as a curious footnote to serious bog archaeology.

*   *   *

Many of the early-discovered bodies were damaged by local people—Yde Girl's teeth were pulled out, her hair ripped away for macabre mementos. Later the machine peat cutters pulverized everything. Many a piece of bread has been toasted over the heat from some old bog man's shreds. Even with best intentions and scientific thoroughness in analyses it is really difficult to understand preliterate people and almost impossible to describe their lives—no matter how much we long to know their stories. The American poet-critic Randall Jarrell (1914–1965), who was labeled "almost brutally serious about literature," chose thirty short stories for a collection, a labor he compared to "starting a zoo in a closet." He remarked that because of the publisher's insistence on brevity the stories could not be "representative." Nor can the brown, folded skins disinterred from the peats of northwest Europe be representative nor tell us the general reasons why or how they came to be there. Jarrell's seeming criterion for inclusion in his book was that each story must present a bitter truth; perhaps the same is true of each bog body.

Quite often in older accounts of the bog bodies I came across statements that the disinterred males found near each other were likely homosexuals punished for flouting presumed prehistoric social norms. I was skeptical. We know nothing of the moral practices and codes of these ancient people, yet this idea seeped into modern folklore, and a belief that prehistoric homosexuals were put to death in bogs still lingers. The idea partly reflects nineteenth- and early twentieth-century moralistic attitudes toward homosexuality bolstered by varied interpretations of two words in Tacitus's *Germania* where the historian, in describing German justice and punishment, refers to human bog remains as *infames corpores*—literally "disreputable bodies." Several different meanings are laid at the door of these words: White's 1866 Latin dictionary offers the meaning "of ill report, disreputable, notorious" for the singular adjective *infamis*. The 2005 *Oxford Desk Dictionary* gives "disreputable" and "infamous." The 2013 revised Oxford translation of *The Germany and the Agricola of Tacitus* translates the plural adjective *infames corpores* as: "Traitors and deserters are hung upon trees: cowards, dastards, and *those guilty of unnatural practices*, are suffocated in mud under a hurdle." (My italics.) Somehow "unnatural practices" crept into the meaning of *infames*. A number of scholars have looked for and found the germ of that idea.

The idea of punishing homosexuals two thousand years ago by drowning them in bogs seems part of the warped Nazi interpretation of Tacitus's essay. The historian Christopher B. Krebs's *A Most Dangerous Book* sees Germany's acceptance of Tacitus's descriptions of the ancient Germans in the 1930s as the blueprint for the propaganda fantasy of national racial purity. Karin Sanders and others link the belief to Himmler's 1937 speech to the Waffen-SS.

> . . . we do not have it as easy as our ancestors did. They only had a few abnormal degenerates. Homosexuals, called Urnings, were drowned in swamps. The worthy professors who find these corpses in the bog are clearly not aware that in ninety-out-of-a-hundred cases they are faced with remains of a homosexual who was drowned in a swamp along with his clothes and everything else.

Following a different route some scholars have picked through other classical historians than Tacitus looking for clues showing that German, Celtic and Gallic warriors were comfortable with homosexuality, but *infames* does not seem related to sexual matters in their works. Livy describes a deserter with the words *infames corpus.* The words could also refer to an unworthy person, meaning a slave. There is certainly the sense of low-valued persons in Tacitus's *infames corpores* but today the Greek and Roman classics writers are seen as primarily reflecting the attitudes of the elite two thousand years ago rather than as eyewitness observers. In the pages of *From Pagan Rome to Byzantium,* the first in the multivolume series *A History of Private Life* by historians of the French Annales group, there is room for a somewhat fuller historical interpretation. Slaves were nonpersons in Tacitus's Rome, though that changed in later centuries.

. . . to allow oneself to be buggered was, for a free man, the height of passivity (*impudicitia*) and lack of self-respect. Pederasty was a minor sin so long as it involved relations between a free man and a slave or person of no account. Jokes about it were common among the people and in the theatre, and people boasted of it in good society. Nearly anyone can enjoy sensual pleasure with a member of the same sex, and pederasty was not at all uncommon in tolerant antiquity.

One has to be careful when assigning motives for presumed actions of millennia-long dead people.

*   *   *

Many fens, bogs and swamps show the human need for a story. How and why did Tollund Man, with a face as unknowable and enigmatic as Mona Lisa's, come to be hanged and put into a bog? What of the two-thousand-year-old Weerdinge Couple found in the Netherlands in 1904? Those two side-by-side bodies seemingly placed in the bog in gentle embrace were popularly called Mr. and Mrs. Peat-Bog. Years after their discovery Mr. and Mrs. Peat-Bog turned out to be two more-or-less headless men and then the story-makers assumed they were a homosexual couple. The larger man's stomach had been slit open and his intestines piled on top of his body.

What of the Windeby "Girl," the subject of Seamus Heaney's poem "Punishment"—". . . her shaved head / like a stubble of black corn / her blindfold a soiled bandage . . ." Discovered in 1952 in a German bog the fourteen-year-old Windeby "Girl" was naked, her hair apparently cut short on one side of her head and she wore a "blindfold." A few yards away was the body of a middle-aged man. The need for a story enflamed imaginations to project an adulterous affair between the two that had led to punishment by death. Heaney's poem follows the adultery path, although one scholar saw "Punishment" as a reflection of the Irish Troubles. Windeby "Girl" crept into many books but in 2007 the biological anthropologist Heather Gill-Robinson, while working on the German Mummy Project in Mannheim, had occasion to examine the remnants of the Windeby "Girl" skeleton (which had been removed from the body during the conservation process) and found the body was that of a sixteen-year-old boy showing traces of malnutrition and/or repeated illnesses. The woolen "blindfold" was likely a band to keep hair in place that had slipped down over the eyes as the body shrank.

The cropped hair on one side was explained as normal decay, for the upper side of the head had been in closer contact with more surface oxygen. And the remains of the presumed adulterous middle-aged lover dated three hundred years earlier than the Windeby Child, as he is now known.

In Ireland the bog body finds were not the same mix of seemingly ordinary people, but were sacrificed kings. Ireland's High Kings go back thousands of years, each with his own kingdom, each conducting warfare with other kings and each responsible for the well-being of his realm. The most famous was Brian Boru. But if the harvests failed or other catastrophe struck the king was sacrificed. Oldcroghan Man, about twenty years old, is believed to be such a royal offering.

> Eamonn Kelly, an expert in Irish bog bodies . . . posits that Oldcroghan Man was a failed king, contender to the throne, or royal hostage sacrificed to a fertility goddess. His carefully manicured fingernails, unworn hands, and last meal of cereals and buttermilk imply high social status. His nipples appear to have been cut, perhaps as an attempt to make him ineligible for the throne. Sucking a king's nipples was a medieval gesture of submission that may have extended as far back as the late Bronze Age . . .

In trying to understand the past we easily make assumptions that seem logical but only reflect our own culture and times, something we recognize in the Roman historians but often not in ourselves. The weight of this baggage is a constant problem which we try to counter with criticism and reflection and the constant reminder that our contemporary orthodoxy of ideas and ways of thinking inevitably cloud every interpretation of why people were interred in bogs.

Steven Mithen, Professor of Early Prehistory at the University of Reading, is one who made the effort to understand early people in *After the Ice: A Global Human History from 20,000 to 5,000 B.C.* Written like an annotated saga Mithen's engrossing book takes the reader on a fantastic prehistoric journey in company with a time traveler, a modern archaeologist yclept John Lubbock who carries the real book *Prehistoric Times*, written by a real Victorian archaeologist—John Lubbock. Two same-named archaeologists and the book within the book confuse until the reader gets the hang of following an invisible and silent guide who is able to move through centuries and millennia and provide the *longue durée* overview. Each of Mithen's interpretations is supported by evidence in a hundred pages of voluminous and rich notes. But one of the most compelling comments Mithen makes is about writing history:

> The challenge we face . . . is not just to combine these [technical reports] sources of evidence so that we can imagine actual communities of plants, animals and insects, but also to gain an understanding of the experience of those who first entered and then became part of those communities. The lists of plants and animals are a poor substitute for the smell of pine needles and the taste of venison roasted under the stars; a report on insect remains cannot evoke the buzz and sting of a horsefly; estimates of winter temperatures fail to convey the numbing pain of fur-clad frozen feet that have walked through snow and waded across icy rivers. Fortunately such sensations are within our grasp: to be a good prehistorian

one should . . . actually go walking and immerse oneself
in the natural world, edging a little closer to the hunter-
gatherer experience.

Early northern Europeans lived and prospered among glacial meltwater wetlands for thousands of years. And what if
those old people who venerated springs, pools and wetlands
as holy places could look into the future and see us? Surely
they would be unable to comprehend humans who dirtied,
drained and destroyed water sources, who dammed and polluted rivers, who choked the great oceans with debris and
plastic. They might well see in us obligatory candidates for
sacrifice to Nerthus.

For some archaeologists the most important peat bog find
was not a prehistoric European body, but the discovery of
Monte Verde in southern Chile which showed (disputably, for
some) that the stone projectile point found near Clovis, New
Mexico, in the 1920s and accepted for fifty years as the indi-
cator of the earliest human settlers in the Americas (11,500 to
10,800 years ago) was predated by at least one earlier arrival.
Monte Verde, preserved under a peat bog that likely grew over
the site after it was abandoned, showed evidence that shat-
tered the myth of "Clovis first." In the 1970s logging along
Chinchihuapi Creek tore up the ground which, like many
damaged logging sites, began to erode. A local family noticed
some bones sticking out of the sagging stream bank. They
collected the bones—presumed to be those of a cow—and
gave them to an acquaintance, a veterinary student studying
at the University of Valdivia. The "cow" bones looked strange,
and the student showed them to several anthropologists at the
university. They came to the attention of the archaeologist

Tom Dillehay, who was teaching at the Universidad Austral de Chile. He was immediately interested. In 1976 Dillehay began a twenty-year excavation of the site, amazed at the extraordinarily excellent preservation of objects reliably dated to 14,800–13,800 BP, "one of the most remarkable archaeological sites in the New World." "Remarkable" because of the excellent condition of the finds. The peat bog had preserved hearths, scraps of skin clothing, a human footprint, fruits and berries, house planks, human coprolites and even a piece of meat whose DNA when analyzed was identified as a gomphothere—an elephant-like creature now extinct and whose protruding bones at the site had been presumed bovine. With the discovery of the Monte Verde bones and artifacts, the earlier belief that the first people in North America had crossed the Bering Strait from Eurasia to North America on foot and left earliest evidence of themselves near Clovis was shaken. Scientists began a serious discussion of alternative or additional coastal boat or raft travel to South America along the resource-rich kelp highway of shoreline. A more vigorous shaking-up came with the 2009 find of ancient footprints in White Sands National Park in New Mexico by David Bustos, the park's resource manager. In the years since he first found the thousands of footprints uncovered by erosion—footprints of adult and child humans, dire wolves, mammoths, camels—originally impressed in the damp earth near the edge of a lake—scientists from many countries have studied the tracks. In 2019 two research geologists, Jeffrey Pigati and Kathleen Springer, noticed the erosion had also uncovered thick layers of seeds of ancient "ditch grass" that had grown near the old lake. Back in the laboratory the seeds were carbon-dated but registered an age so great—22,800 years ago—the geologists designed a series of rigorous tests to verify the startling dates if the seeds were contemporaneous with the footprints. As of now the seeds and the footprints are accepted as the earliest proof of humans in the Americas. The humans, whether they came by land or by sea, were the ancestors of most of the indigenous cultures in the Americas.

Das Grosses Moor at Kalkriese Gap in AD 9

Sometimes almost-forgotten ancient events come back to life with new importance. A longing for inclusive history tinctured nineteenth-century Germany in the years around the country's 1871 unification. Because Germany has loomed large in world history for the last hundred and forty years it is easy to forget that for a thousand years following the breakup of Charlemagne's empire "Germany" was a maddeningly complex Grimm's fairy tale jumble of small states ruled by individual princes and kings, small states that shared the bond of a common language and expressed a sense of unity through coalitions and economic collaboration. After German victory in the 1870–1871 Franco-Prussian war, Prussia, the militarily strongest state, took a leadership role and in 1871 the Treaty of Versailles confirmed the legal harmony of German unification.

The new nation had a psychological need to show some proof of its historical greatness. The most shining example was the brilliant victory of AD 9 when Arminius of the Cherusci tribespeople led a consortium of indigenous farmer-warriors in annihilating three Roman legions—the so-called Battle of the Teutoburg Forest. The battle is still meaningful and some believe the humiliating defeat of Rome's finest by illiterate "barbarians" froze the Rhine into a permanent boundary dividing eastern and western Europe.

We may think of Roman legionnaires as big tall men, but years of archaeological measuring of Roman bone lengths discloses that the men were an average height of five feet five. In contrast, the kingly bog body of Oldcroghan Man was six feet three. But indisputably the professional Roman army was the most formidable in the world. The legions were made up of highly trained (however short) men wearing armor, hobnailed sandals and brilliant red cloaks; in close formation they could march twenty miles in five hours. (Probably some could sleep as they marched, as Tobias Wolff's character Wingfield does.) They wore iron helmets, carried *framea* spears, javelins, a sword and a last-resort dagger as well as water bottles and cooking pots. They wore amulets and rings. Their weapons were rich with silver and gold. Each man had a painted shield with which in a unison sweep he could form a protective wall and roof—the famous *testudo*. They had a reputation for brutality. The taller and heavier German farmer-warriors wore trousers and tunics of wool and linen. Each had a dagger and carried a spear. What they lacked in weaponry they made up for with intimate knowledge of the terrain of their boggy, marshy homeland, and the tactical use of bogs and swamps in battle.

The freshly unified Germans considered Arminius (centuries later Germanified to "Hermann" by his admirer Martin Luther) a national ur-hero who deserved a monument. The correct place for it would be the battle site—wherever that was. The specific location was unknown. The only clue was a sentence in Tacitus's *Germania* that named "Saltus Teutoburgensis" as the battlefield. The German military historian Hans Delbrück (1848–1929) believed this meant the Teutoburg Forest near Detmold.

A different suggestion came from the historian Theodor Mommsen (1817–1903), an authority on ancient Rome; his attention had settled on finds of Roman coins dating from the reign of Augustus discovered a few years earlier by locals near Osnabrück and Kalkriese Hill in Lower Saxony. Mommsen believed the coins came from the three destroyed legions. He also noticed that the area was ideal for an ambush. Yet the authors of the Greek and Roman classics were indisputable authorities, and the leading German historians ignored Mommsen in favor of the ancient Roman historian Tacitus, who had not been born when the battle occurred, never went to the site and wrote his account ninety years after the event. Yet the need for a hero resulted in the installation of a 175-foot-tall monument of Arminius/Hermann—*Hermannsdenkmal*—in Detmold. It went on public view in 1875 to immediate popularity.

Modern scholars reexamining Tacitus think that the word *Saltus* did not mean forest (*wald*) but "pass" or "gap"—a narrow bottleneck—and they cite Livy's use of *Saltus* in his description of the pass in the Battle of Thermopylae (*Thermopylarum Saltum ubi angustae fauces coartant iter*). In the long run the historian Mommsen had been correct. The Battle of Teutoburg Forest took place in Lower Saxony on a constricted track between wooded Kalkriese Hill and *das Grosses Moor* or Great Bog—more than fifty miles distant from Detmold. A more accurate name for the ambush might have been the Battle of Kalkriese Gap. An ambush in a narrow defile was an ancient hunting technique in the region, as Steven Mithen describes with archaeological evidence of hunters slaughtering reindeer in their annual migration through the Ahrensburg Valley in Schleswig-Holstein 12,600 BC.

The discovery of the true site came in 1987 when Lieutenant Tony Clunn (1946–2014) of the British Army of the Rhine was stationed at Osnabrück. Clunn was interested in numismatics and his hobby was searching for coins with a metal detector. Before poking into the local corners he visited the regional archaeologist Wolfgang Schlüter at the Osnaburg museum and asked permission to look for Roman coins and for suggestions of likely places. The archaeologist was "very cautious," commenting that no Roman coins had been found in the area for years, but he suggested a place to try.

Clunn's 2005 book *The Quest for the Lost Roman Legions* is his personal account and shows his experienced eye for terrain and topography. On his first day out he examined a field and noticed

> . . . a slight elevation running across the field, possibly part of an old track or trail . . . as I neared the center point of the track I heard a familiar double-ringed tone in my headset . . . I cut away a square of turf . . . continued carefully to clear out the black peat within the hole . . . picked up a handful of soil . . . sifted through the contents in my hand . . . sifted through again and then I saw it: black, small . . . and round! . . . it was a perfect silver coin . . . a Roman denarius.

Dr. Schlüter was away on vacation. Over the next few weeks Clunn recovered more than a hundred Roman coins, none later than the crucial Augustan years. When Dr. Schlüter finally returned Clunn brought in his envelope of coins. The stiff and formal archaeologist unbent and the two men gradually became good friends, a friendship that endured during years of Schlüter's excavations and reports on the site of the *Varusschlacht*. Clunn continued to study old maps of the region to find trackways and faded roads. The search brought him closer to the area at the base of Kalkriese Hill, in modern times dry farmland but sopping wet bog two thousand years earlier. Most accounts of the battle ignore *das Grosses Moor*, but the peat in which Clunn found the coins indicates the agricultural field had once been a peat bog.

Clunn described one of his early finds.

. . . something I had recovered earlier and handed in unrecognized with other metallic material turned out to be a find of great significance. Three separate lead ovals generated a tremendous level of excitement with Dr. Schlüter and the museum staff. He quickly recognized them for what they were: lead slingshot—the first real evidence of a military presence or engagement (or even a battle) in the Kalkriese area.

Modern trials prove that almond-shaped lead shot in the hands of an experienced slinger could be accurate with a range of 130 yards and nearly the force of a .44 magnum bullet. When grooved, the missiles made an eerie sound as they sped through the air. Soldiers often scratched images of scorpions or rude messages—"here's a sugar plum for you"—on the shot before using them. The German warriors did not use slingshots, and so finding them at Kalkriese was proof that Roman soldiers had been there.

Rome's first emperor, Caesar Augustus, was seventy-two years old and near the end of his rule when the legions suffered their catastrophic defeat on the edge of the Great Bog. Germania's population was rural, made up of farmer-warriors and their families living in small settlements at the time of the battle. There were no real towns, and private ownership of land had been unknown among the eastern barbarians fifty years earlier when Caesar conquered Gaul. In general, when colonial- and imperial-minded aggressors make their moves into new territories and encounter indigenous people, often very numerous and "complex, multi-lingual, culturally diverse," as the two groups gradually mix and confront each other, tribal identities begin to take shape and individual "tribal leaders" are named. For the aggressor, this bundling is the opening process of controlling the indigenous people who, up to that time, may not have seen themselves as distinct tribes. Suddenly they are corralled by identity to a specific area.

The Roman system of conquest was to grant conquered people Roman citizenship and involve them in Roman customs and culture. What Rome got from its aggressive takeovers encircling the Mediterranean Sea was an increase of manpower to serve in the army, slaves and money from taxation of its new colonies.

The Roman legions were augmented by auxiliaries of men from conquered lands. Yet many of the vanquished hated the Romans, their martial ways, their enslavements, their self-proclaimed superiority, their heavy taxes and their strutting presence as overseers and governors in seized territories. At the same time the conquered population wanted to be joined to the powerful, to visit glittering Rome whence all roads led.

After his conquest of Gaul Julius Caesar had had plans to take on the farmer-warriors east of the Rhine but he was assassinated in 44 BC. His adopted heir, Augustus, succeeded him. In 11 BC Augustus's adopted son Drusus marched as far east as the Weser River, home region of the Cherusci. His troops had a hard time. Two years later when he tried again, he made it as far as the Elbe, then he fell from his horse, broke his leg and eventually died from gangrene. In Rome his feat of reaching the Elbe was celebrated as a victory based on the erroneous assumption that Germania had been conquered. There *had* been a battle, and according to some accounts Drusus had defeated the Cheruscian nobleman Segimer, and, in the practice of that time, he took Segimer's oldest son, Arminius, as hostage to guarantee his father's good behavior and sent him to Rome.

In AD 4 and 5 Drusus's brother Tiberius campaigned as far as the Elbe without claiming he had conquered the territory, yet Augustus believed the region was now ready to become a Roman province. He appointed the general Publius Quinctilius Varus to the post of legate and governor of this presumably surrendered region. It was Varus's job to set up the standard Roman administration and tax collection. The way he went about it led the historian Simon Schama to describe Varus as "the Custer of the Teutoburg Forest," a man exuding "racial and cultural arrogance."

Young Arminius learned to speak and read Latin and enlisted in the Roman army, leader of an auxiliary unit. He received the award of Roman citizenship given to superior auxiliary leaders. "Superior" might have reflected his high Cheruscian social rank rather than outstanding performance. If he served Rome during the Pannonian campaign an encounter at Sirmium (near modern Belgrade) might have caught his attention. A surprise attack by locals in a swamp nearly cost Romans the battle. Perhaps Arminius/Hermann remembered a familiar bog-and-forest setting in his native Germania. He was back home again when Varus assumed the governorship. He dined with Varus, smiling while plotting treachery. He was so friendly that when his father-in-law, Segestes, an ally of Rome who disliked Arminius for marrying his daughter without his consent, tried to tell Varus that Arminius was plotting something, the governor didn't believe him.

The usual way armies fought in those days was on open ground where opposing forces lined up facing each other, then marched forward until they met in hand-to-hand combat. The Kalkriese battle was different: a surprise ambush, a forested slope, the edge of a vast bog, its dark water exhaling mist, and a chokingly narrow combat area.

Arminius's ruse was a false report to Varus that a local tribe was causing trouble and needed attention. Varus did not doubt. On the day of the battle as the legions marched along the shrinking pathway toward the imaginary tribe they were increasingly crowded together, treading on one another's heels. To step off the pathway was to step into the mud-sucking bog on the right or the sloped woodland on the left.

Arminius's thousands of men, behind the trees and behind a turf wall, let many of the Romans pass. They waited, waited and then suddenly attacked the bunched-up troops. Horses, mules and men stumbled and fell on the path or fled into the reddening bog water. In only a few minutes thousands of Roman soldiers fell dead or wounded, streaming blood. Arminius's men rushed down to continue the massacre. If each man lost three of the nine to eleven pints of blood in his body the pathway and bog would have been flooded with thousands of gallons of blood. When Varus heard, he and his officers did the only honorable thing—they fell on their swords. Over three days roughly 13,000 to 16,000 Romans died and about 500 Germans. The Germanic men captured more than 1,000 Roman soldiers and later ritually killed them on sacrificial altars, or saved them to offer to the sacred Great Bog that had given them their victory. Some were hung from sacred oaks in the forest. The victors nailed Roman heads onto trees, gathered up Roman weapons and equipment, some kept and used, many weapons, coins, amulets, bells and a silver-plated parade mask all flung in votive thanks to the avid water gods of the bog. They could hear the black bog waters glugging as they accepted the offerings.

Far distant in time and space from Arminius's Great Bog I walked along the Strait of Juan de Fuca shoreline. I thought the layered peat bed resembled moldy lasagna. Above it rose the bluff, its khaki soil stuffed with rocks and bones, souvenirs of its journeys as the ice lobes pushed the dirt around, advancing, retreating. Streams and rivers of ice-melt water built up the bluffs; the inevitable erosion began courtesy of Puget Sound waters and storms.

In some earlier wet year the bluff rearranged itself with an earth slide that settled into alternating terraces and slopes. The spillover on the beach was carried away by storms and tides that licked the face of the peat bed clean and polished the boulders—prizes in a geologic Cracker Jack box. The bluff's new arrangement in a descending series of plateaus was claimed by *Equisetum telmateia*. I imagined that the narrow boggy path where the German farmer-warriors fought the Romans was edged with masses of great horsetail, a plant descended from giant ancestors first seen in the Devonian that somehow survived the next five mass extinctions. The range of *E. telmateia* is from Sweden south to Germany, east to North Africa and Asia and along the west coast of North America. In earliest spring it sends up potent strobili whose spore-packed cones balance atop pinkish tan stalks striped with fringed black bands like miniature grass skirts. In summer the bright hollow stalks of the horsetails, segmented like bamboo, crowd the slope. Their vibrating color is the fresh flash of young poplar leaves. Their stem joints show distinctive bands of black, antique white and pale green, and from each joint there emerges a whorl of segmented "leaves," each the size and shape of a slightly cooked strand of spaghettini. The entire plant is rich in silica and has been used through the ages as fine sandpaper. *Equisetum telmateia* was the ancestral food of dinosaurs and today bears coming out of hibernation still seek out the plant (the silica content is a stern diuretic) to get their digestive systems working. Throughout the summer the plants remained brilliant towers of irradiant green, shoulder-high to humans, but in autumn they suddenly fainted and collapsed. In winter I walked past the fallen horsetails, their lively color leached to the tones of dead flesh, the joints still demarcated by grey bands so that from a distance their sprawl across the slope resembled discarded lines of code.

Something ended on this last walk and on the way back to my house I took the decision to leave the Pacific Northwest and return to New England after an absence of twenty-five years. It would be like going to another country. It would be like time travel. There would be different horsetails.

4.

SWAMP

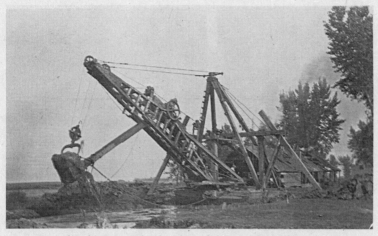

Dredge boat on the banks of the Kankakee at work

In 1922 Akutagawa Ryūnosuke (1892–1927) wrote "In a Grove," a disturbing story that packed into a few pages the twisted ways humans perceive and report personal experiences. Twenty-eight years later the story became *Rashomon* with the 1950 release of Akira Kurosawa's film adaptation. The story presented conflicting and confusing versions of a killing in a forest. Akutagawa took the germ of the story from a twelfth-century collection of more than a thousand ancient tales from China, India and Japan, the *Konjaku Monogatari*, or *Tales Old and New*, a source he often used for story ideas. "In a Grove" seems made for times of change, upheaval and violent events when falsehood and legerdemain obscure every reality in life, when people seek the prophecies of seers, the sordid enchantments of amoral magicians and—without reflection—assault the natural world.

These peatland pages have cast a glance at England's great fenlands and at northern Europe's historic bogs and the human uses and ecological violence visited on them. The third peatland is the swamp the despicable, exquisite, confounding, ever-changing swamp. Over the last four centuries Americans have manhandled the country's peatlands, but the personalities and names of the Great Dismal Swamp, Black

Swamp, Okefenokee, Everglades, Kankakee, Limberlost and their kind still resonate with meaning.

A swamp is a peat-forming wetland that can support shrubs and trees—even forests—as the cypress and tupelo swamps of the American South. A swamp is distinguished by its woody character. North American hardwood swamps are red maple, black willow, aspen, cottonwood, ash, elm, swamp white oak, birch and black gum. Coniferous evergreen swamps are habitat for Atlantic white cedar, black spruce and balsam fir. The deciduous larch lives in many northern swamps where its brilliant yellow needles glow like torches in autumn.

Wooded swamps are at the end of the fen-bog-swamp succession, just one step from becoming dry forest. They are legacies of the ice age when the melt started the sequence by creating stupendously huge lakes—Lake Agassiz covered 170,000 square miles of Manitoba, Ontario, North Dakota, Minnesota and Saskatchewan. Lake Missoula covered about 3,000 square miles of what is now Montana, Washington and Idaho. A century ago the strange terrain and huge current ripples caught the eye and the interest of a geologist at the University of Washington. In the 1920s after considerable map study and ground observation J Harlen Bretz stepped off the reservation of gradualism to present his theory that Missoula's repeatedly bursting ice dams, its cataclysmic floods created the bizarre giant earth ripples as the long-ago gushers scoured out the channeled scablands of eastern Washington. Not gradualism but sudden and catastrophic meltwater floods had blasted out coulees, hollows, and the monstrous three-mile gaping gorge of the Columbia. Decades passed before his heretical evidence was grudgingly accepted.

The original occupants of the continent knew the rivers and swamps, the bogs and lakes as they knew the terrain and each other. But for most English settlers and European new-comers to what was later the United States, nature consisted of passive and inanimate substances and situations waiting to be used to human advantage. Preservation and care of nature was not what they had come for. As the country grew its preference was arable farmland, not wetland.

In the nineteenth century the United States enlarged itself in a fever of land acquisition: the 1803 Louisiana Purchase of 800,000 square miles from the Gulf of Mexico to Canada doubled the size of the country; in 1819 the Adams-Onis Treaty added Florida and part of Oregon; 525,000 square miles of Texas was annexed in 1845; the Oregon Compro-mise in 1846 enrolled the Pacific Northwest from northern California to the Strait of Juan de Fuca. Great oceans and lakes framed the country and the interior roiled with tangles of rivers like unspooling silver ribbons. All that territory had once held a wealth of wetness—scientists have estimated ap-proximately 221 million sopping acres existed in the early seventeenth century, much of it swamp. As the United States pushed its borders, thereby gaining wetlands, its population leaped from 7.2 million people in 1810 to an astounding 12.8 million by 1832, almost doubling in twenty-two years. The welcoming arms of open immigration became the hallmark of America and that reputation lingers in global memory de-spite today's more painfully stringent reality.

During the Civil War moving heavy guns and personnel through swamps was arduously difficult; soldiers resorted to the time-consuming work of clearing bypass routes. One of the men later wrote that after the First Gum Swamp battle wading through knee-deep and deeper mud: "The brambles [were] thick and thorny, the water coffee-colored, alive with creeping things, the air heavy with moisture and foul odors." These memories lingered. As the country grew the ongoing stories of vile adventures in the muck made it clear to military, government and citizenry that something had to be done about the swamps so universally detested. Everywhere there were horrendous mixtures of fen, bog, swamp, river, pond, lake and human frustration. This was a country of rich absorbent wetlands that increasingly no one wanted and most despised.

Swamps are often (erroneously) called marshes, but the distinguishing difference is that marsh vegetation is usually grasses and reeds; swamps are woody shrubs and trees. In practice all wetland labels are freely bandied about in conversation, books, reports, literature, media. Marjory Stoneman Douglas called the Everglades "a river of grass," Michael Grunwald called it a "swamp," Indiana settlers delighted in the bountiful Grand Kankakee Marsh, Ohio people cursed the Great Black Swamp until they made it into Great Black Farmland, and Canto 7 of Dante's *Inferno* describes Dante and Virgil following a stream through a deep gully descending into the Fifth Circle of Hell. Dante did not mention grasses or reeds in the marsh, but he did specify its origin in a high-ground spring that became a vigorous stream pooling in the fifth circle of hell. This literary wetland would seem to be minerotrophic—a non-peat-forming marsh—while throughout the lower reaches of hell there were smoldering zombie fires of peatlands that will burn underground forever.

And when that dismal stream had reached the foot
Of the malign and dusky precipice,
It spread into a marsh that men call Styx.

John Stilgoe, in his discussion of estuary English, writes of the
mix-up of names for passageways through the salt marshes
where the words *creeks*, *brooks*, *ditches*, *gutters*, *guzzles* are
thrown about indiscriminately. Ponds, lakes, holes have their
own intricate variations. It is unlikely this water-word con-
fusion will ever clear up any more than we will ever know
what happened in Akutagawa's forest. Because the peatlands
are so variable and in constant change, often embracing all
stages of bogs, fens and swamps within a single perimeter, for
the ordinary person choosing the correct term is sometimes
a toss-up, especially as wetland vocabularies vary from place
to place. It is like that other ancient story from the Indian
subcontinent—the parable of the elephant explained by unknowing blind men. The story has spread the world over and
is embedded in Buddhist, Hindu and Jain texts. Each man
touches a different part of the elephant and concludes that
the beast is very like a snake, a fan, a wall, a spear. It is an
example of alternate and partial truths that must be combined to arrive at a holistic truth. Peatlands are a little like the
elephant and putting a label on one can depend on what part
of the elephant you touch. We see what we see, not necessarily what is there, and we write what we know, not necessarily
what we see.

It can be hell finding one's way across an extensive boggy
moor—the partially dry rough ground and the absence of
any landmarks lets the eye rove helplessly into the mono-
type distance. Everything undulates, the rise and fall share

the same muted palette and the senses dull. But a swamp is different. Although water and squelch are everywhere, there are landmarks—downed trees or jagged stumps, a tenanted heron nest, occasional islands of high-ground hardwood stands called "hammocks" in the south. Yet the swamp traveler goes not in a straight line, but slouches from quaking island to thick tussock to slippery half-submerged log. Even with GPS technology big swamps are places to get lost, and in the past many who had reason to melt out of sight—native people threatened out of their territory, runaway slaves, Civil War army deserters, moonshiners and bloody-handed murderers—have hidden in them. For a few minutes I once considered hiding in a swamp myself.

When I was ten years old my family lived in a rented house in Rhode Island. In my memory its distinguishing feature was a large angular hole in the wall of the lower stair landing. The hole was the shape of a human arm. We were temporary inhabitants and paid little attention to it though in retrospect it was evidence that someone had hurtled down the stairs and hit the wall with force. After the hateful school week Saturdays were free time and I sometimes went on my own to a nearby swamp circled by a fisherman's path. Far out in the water stood the unreachable hulk of a dead tree—branchless, tall and sun-bleached white with a large hole near the top. I had somewhere read that great blue herons nested in such snags and that in one swamp a man had brought a ladder, placed it against the tree and climbed up to look into the heron's nest. The heron had stabbed him in the eye as he came level with the nest and the man, his eye and brain pierced, fell dead from the ladder. I wanted to see if there was a heron nest in this local swamp's dead tree—

perhaps even a live heron—perhaps even the remains of a ladder—perhaps even a sun-bleached skull on the ground. When I got to the swamp there was a small raft and pole lying on the bank. I had never seen one there before. There was no one around. Perhaps it was abandoned. Yes, likely it was abandoned. It was my chance. I pushed it out into the tawny water, got on board and began poling toward the snag. I was halfway there when I heard furious shouts and screams. Looking back I saw the two worst boys in the school jumping up and down on the bank and hurling futile clods of mud. I had stolen their raft. After a quick look for a nonexistent hiding place I changed direction and took an oblique route to the farthest end of the swamp where I pole-vaulted onto firm land, found the path and rushed away from the scene of the crime. It was some time before I noticed I was still carrying the raft-pole and I leaned it helpfully against a tree before continuing home.

Many modern Americans do not like swamps, herons or no herons, and experience discomfort, irritation, bewilderment and frustration when coaxed or forced into one except for a few (like my mother) for whom entering a swamp was to plunge into a complex world of rare novelties and eerie beauty. One of my mother's heroes was Henry David Thoreau (1817–1862), the enigmatic New England surveyor-naturalist-essayist. Thoreau has been called the patron saint of swamps because in them he found the deepest kind of beauty and interest. He wrote of his fondness for swamps throughout his life, most feelingly in his "Walking" essay:

> Yes, though you may think me perverse, if it were proposed to me to dwell in the neighborhood of the most

beautiful garden that ever human art contrived, or else a dismal swamp, I should certainly decide for the swamp.

He went so far as to describe his dream house with windows fronting on a swamp where he could see "the high blueberry, panicled Andromeda, lamb-kill, azalea, and rhodora—all standing in the quaking sphagnum." (This would seem to be a transitional stage between bog and swamp.)

By the 1980s American wetlands were wiped out roughly by half, in some states considerably more. Aerial photography made wetland size estimates possible and in 1990 the U.S. Fish and Wildlife Service published a study showing that since the 1600s the country's treasury of wetlands had shriveled to 103 million acres and some states had lost almost all of their original wetlands. Then, between 2004 and 2009, another 62,300 acres disappeared to agricultural interests and housing developers. They continue to disappear through sediment deposition patterns, fertilizer runoff, spilled and leaking chemicals, increasing floods, storms, droughts, fires and today's rising sea level.

Rising sea level is both subtle and blatant—we hardly notice it until a storm brings vast flooding. The Mesolithic people had no cities and a small population—they had room to move around; we have huge cities and not much unoccupied livable space. A popular example of our rising water level problem is Naval Station Norfolk in the Hampton Roads, a natural roadstead channel of deep water in Chesapeake Bay fed by the James, Nansemond and Elizabeth rivers, where now sea water is swelling up at twice the rate elsewhere. The environmental writer Jeff Goodell visited the station and wrote, "There is no high ground on the base, nowhere to retreat to. It feels like a swamp that has been dredged and paved over—and that's pretty much what it is." Although the danger for this one base is well known, little remedial work has been done thanks to bureaucratic red tape, pockets of climate denial in the House of Representatives and the strange illusion that there will be time to fix the problem.

After a rainstorm any curious child who drags a stick obliquely and deeply away from a rivulet sees the rivulet forsake its original channel and follow the stick line; the stick dragger has discovered the principle of drainage. This innate existential curiosity has led humans to commit unthinking malfeasances against the natural world. Farmers grew up with shovel in hand ready to cut drainage ditches. Government was solidly on the side of drainage to increase land area, in part for incoming immigrants. In 1849 Congress passed the first of several Swamp Land laws that turned federal wetlands over to the individual states with the right to dispense those water-sodden acreages for purposes of drainage. These laws perpetuated the myth of endless land free for the taking, and showed an inability or unwillingness to observe changes in nature over the seasons and years.

Yet there was some cognizance of destruction, especially in the opinions of the Vermont statesman-farmer George P. Marsh. His 1874 *The Earth as Modified by Human Action* was a century before its time. He wrote:

> . . . man is everywhere a disturbing agent. Wherever he
> plants his foot, the harmonies of nature are turned to
> discords. The proportions and accommodations which

insured the stability of existing arrangements are over-thrown. Indigenous animal and vegetable species are extirpated, and supplanted by others of foreign origin, spontaneous production is forbidden or restricted, and the face of the earth is laid bare or covered with a new and reluctant growth of vegetable forms, and with alien forms of animal life.

In the early twentieth century the annoying disappearance of waterbirds was charged to the ravages of market gunners who in every month of the year slaughtered at will. The stunning carnage of the millinery plume trade killed as many as 5 million birds a year. Michael Grunwald, describing the black comedy of the destruction of the Florida Everglades, wrote:

> At the height of the nesting season, plumers patiently shot out rookeries one bird at a time, leaving rotting carcasses and helpless chicks to be devoured by raccoons, crows and buzzards. They used quiet weapons . . . so their shots sounded like snapping twigs. The birds barely noticed them, and when they did, the adults rarely left their nests for fear of abandoning their young.

Although the accumulating evidence pointed to swamp habitat destruction as the major cause of waterfowl disappearance, in 1945 the Fish and Wildlife Service published a popular wild game cookery book with a foreword by the famous Republican cartoonist-conservationist "Ding" Darling that still pointed the finger at the market hunters of yesteryear:

By the end of the Civil War, killing for the market had become a serious factor in the reduction of game. Traffic in game reached its peak probably in the 1880's. . . . By every conceivable method, waterfowl were killed by the million and sold in the open markets for a few cents apiece. Old-time market hunters used swivel or punt guns that could kill as many as 50 to 100 waterfowl with one shot.

Recreational waterfowl hunting had gradually become a pastime that attracted legal and medical professionals, wealthy businessmen, judges and politicos—people with voices and opinions that were heard. Over the years they observed that the great flocks of ducks that made hunting so rewarding had—somehow—disappeared. As usual the villains of the blame game were market hunters, yet for the first time biologists also pointed out that draining the midwest swamps to make farmland had markedly reduced the ducks' habitats. The new idea of wetland conservation floated through the air on downy duck feathers.

The great coastal southern swamps of the United States were and are treasures of the natural world. Some have been exploited and damaged beyond recognition, some are still rich and wonderful, preserved as wildlife refugia or parks. Visitors can share the amazement and delight of the botanist William Bartram, whose exploratory travels in Georgia and Florida from 1739–1773 showed a wild tropical south we can know only through his words and drawings—touchily sensitive Seminoles, crafty alligators, exquisite unnamed flowers, masses of bayonet-like grasses, colossus black oaks and anonymous plants.

William Bartram was the botanist-traveler-writer son of the Philadelphia Quaker John Bartram (1699–1777), who had been appointed Botanist for the American Colonies by George III. John Bartram made the country's first botanical garden on his Philadelphia property. Father and son often went on botanical expeditions together. One such was Georgia's Lower Altamaha, where the Bartrams in 1765 first discovered the Franklinia in a sandhill bog. This small beautiful tree is now extinct in the wild, but continues to delight American gardeners who grow specimens all descended from those few seeds collected by William Bartram on his Georgia travels. Thinking of the Bartrams I once planted the closely related Stewartia in my garden in Port Townsend, Washington; it grew handsomely but did not flower while I lived there.

A valuable medicinal plant was Bartram's second find.

On drawing near the fort, I was greatly delighted at the appearance of two new beautiful shrubs, in all their blooming graces. One of them appeared to be a species of Gordonia [*Franklinia alatahama* (*sic*)], but the flowers are larger and more fragrant . . . The other was equally distinguished for beauty and singularity; it grows twelve or fifteen feet high . . . with large panicles of pale blue tubular flowers, speckled on the inside with crimson . . .

This was *Pinckneya pubens*, the Georgia "fever tree," a natural source of quinine used medicinally by native Americans for tick fever, muscle cramps, parasites, fevers and malaria.

On his travels William Bartram remarked that turkeys and bears that dined on oranges and acorns "are made extremely fat and delicious from their feeding." Other times the travels were dangerous or pestiferous, as when he fell asleep next to his campfire to ". . . enjoy only a few moments, when I was awakened and greatly surprised, by the terrifying screams of Owls in the deep swamps around me . . . which increased and spread every way for miles around, in dreadful peals vibrating through the dark extensive forests." This past spring I heard lovelorn owls similarly whooping and caterwauling in the New Hampshire woods.

Brooke Meanley's Swamp Sampler

The American biologist-ornithologist Brooke Meanley (1915–2007) knew intimately every swamp corner the Bartrams had visited two centuries earlier. He spent his professional life in the southern swamps. His observational skills and photographs show what these wetlands were like seventy years ago. Born in Maryland and educated at the University of Maryland, Meanley worked as an ornithologist for the Department of the Interior. In his work he took thousands of pictures of swamp habitats and birds—many places and birds that no longer exist. His book *The Patuxent River Wildrice Marsh* shows the importance of long, continued observation in the decades of changes he saw along Maryland's Patuxent River. He wrote in 1993, "I recall in the early years when I visited the marsh, some 60 years ago . . . a pure stand of Wildrice. Today, much of that same area is a shrub swamp, slowly succeeding to an early woodland stage."

During World War II he served four years stationed in Georgia rehabilitating body- and psyche-damaged returning soldiers. His way was to take the shaking and jittery men on hikes and bird walks through nearby forests and swamps. One can only guess how many bird-watchers and amateur naturalists found mental balance and lifelong interests in the natural world through these expeditions. Possibly they learned from him that cutting old-growth forests removed vital bird habitat.

In his decades of field research Meanley built up a huge bank of knowledge about certain migratory birds—redwing blackbirds, grackles and cowbirds—that battened on the southern rice fields. Rice farmers believed the birds were taking a great part of the harvest. The scientific work of Meanley and others in banding the birds, counting the contents of thousands of bird crops, measuring and observing times of harvest, migration times, planting schedules, ways to trap, kill or frighten birds away and countless other details determined that the depredations were far less than farmers' guesses and that planting and harvesting schedules could be timed to avoid heavy losses. In addition the researchers found that the birds ate considerable numbers of insects harmful to rice. Sometimes it seems that his farm-help conclusions and private feelings were at odds, as when he mildly remarked that brush and woodland around the borders of rice fields made good habitat for wildlife but also were attractive nesting and resting places for blackbirds in spite of the goal of "clean farming" that eliminated brush and trees near the fields to reduce crop damage. "Clean farming" is a little like clear-cutting an old forest: it removes animal, bird and insect habitat. We often get the sense that although he was rigorously doing his job his heart was with the blackbirds. And his eye was on other birds in the rice lands. On regular visits to the blackbird roosts he noticed that in addition to himself predatory mammals and birds monitored the roost areas. "At a typical roost near Hazen, Prairie County, Ark., 79 predatory birds were observed on January 20, 1953, including 74 hawks of three species and 5 owls of three species."

Meanley's years in and around the southern waterlands are encapsulated in his *Swamps, River Bottoms & Canebrakes*. I had never heard of the Slovak Thicket until I read Meanley's description: "For its size, the fourteen acre Slovac Thicket, located in the heart of the Grand Prairie near Stuttgart, Arkansas, packed the most wildlife excitement per acre that I have ever known." It's a good bet that a sky totally black with the 20 million birds that he saw and photographed that day cannot now be seen.

Swamps and birds go together; when the swamp disappears so do the birds. The New World warblers (aka "wood warblers"), a group of about fifty small (smaller than sparrows) passerine birds that migrate from South and Central America to the boreal forests of Alaska and Canada, were Meanley's favorites. Many are brightly colored and their complicated high-pitched songs are difficult to hear. They flicker and flit through branches and reeds like sunlight on a windy day and are a challenge to see. In a perfect world a warbler can live for a decade, but in the world of predatory housecats, wind turbines and enormous glass buildings a warbler is lucky to live two years. Meanley found the bottomlands of the I'On Swamp in South Carolina was a choice habitat for Bachman's warbler, once the seventh-most-common migratory bird, annually up from Cuba to breed in the blackberry swamps and cane thickets of the southeast United States. The I'On Swamp (named for the early landowner Jacob I'On) was the hunting ground for an early American ornithologist, the Reverend John Bachman who in 1833 first found this country's now-rarest songbird—Bachman's warbler. His friend Audubon listed the warbler in his *Ornithological Biography*. As other regions were drained and cut, warblers found a refuge in the I'On. Meanley counted himself fortunate to have twice seen Bachman's warbler in his lifetime—in 1958 and 1963. In his day he knew it was near extinction. It has not been seen since 1977 and is now presumed to have joined the passenger pigeon and the ivory-bill.

The lower White River Wilderness was named a national wildlife refuge in 1935 and it remains Arkansas's dearest wild bird sanctuary. Meanley worked in the Wilderness in the early 1950s always on watch for the famous ivory-bill though he thought it had likely been wiped out in the 1930s. Still, he notes that the ornithologist George Lowery in his *Louisiana Birds* says the last authenticated report for the woodpecker was in 1943 in the Singer Tract, an old-growth forest in northeast Louisiana then owned by the Singer Sewing Machine Company.

The finest hardwood trees grew in the 120-square-mile Singer Tract swamp in northeast Louisiana, described by James Tanner of ivory-bill fame as "virgin forest." Meanley remarks in 1972 that he knows of "no swamp or bottomland forests in the southern part of the United States that as recently as thirty-five years ago contained as many rare animal forms. In addition to the Ivory-billed woodpecker, the Bachman's warbler . . . the panther . . . and the red wolf . . ."

The Singer Tract today is a National Wildlife Refuge in the upper basin of the Tensas River. In the 1930s, before the Tract was cut, perhaps seven pairs of ivory-bills lived there, making their livings on wood-borer beetles under the bark of ancient, decaying trees as Tanner claimed, but also, as three preserved ivory-bill stomachs show, a more varied diet of pecans, poison ivy berries, magnolia seeds and hickory nuts.

Born in 1912 in the Finger Lakes region of New York, James T. Tanner studied at Cornell University under Arthur Allen, the founder of the famed Ornithology Lab. One of Allen's interests was the ivory-billed woodpecker, believed to be extinct until Allen saw a Florida pair in 1924. In 1932 ivory-bills were seen in Louisiana and Allen began a serious hunt for the rare birds with a bird artist, a sound engineer and a graduate student—James Tanner. Later Tanner spent three years studying one family of ivory-bills, now the only biological study we have. Tanner was not without prejudices. Current hopeful searchers for the elusive bird point out that since Tanner was convinced the ivory-bills were old-growth forest obligates he overlooked younger forests as possible refugia. And some are of the opinion that no one man—Tanner— could have examined all the likely swamp forest land that might have supported ivory-bills. Before the loggers and ornithologists came on the scene the indigenous people of the swamps had hunted ivory-bills for their ornamental beaks. Once the bird's rarity was known the professional shooters for museums anxious to fill out their collections—without a thought for living populations—joined the mix of those who unwittingly wrote the ivory-bill's epitaph. In September 2021 the U. S. Fish and Wildlife Service officially listed the ivory-bill woodpecker and twenty more species extinct.

For Meanley, the prince of southern swamps was the Okefenokee with up to twenty-five feet of peat deposits, once a haunt of the ivory-bill. In describing the swamp's charms he wrote that it had everything: "the live oak hammocks, alligators and large wading birds, and the legends. In my judgement it is the most picturesque swamp in North America." It was, he writes, a mosaic of lakes, shrub bogs, cypress heads and cypress bays, and though much of its cypress had been cut in the early twentieth century, fifty years later when he was in the Okefenokee lusty regrowth allowed him to say the swamp "looks today as it did when it was the stronghold of the Seminoles and Creeks."

In the 1950s my then-husband and I sometimes vacationed on one of the Georgia islands—St. Simons or Sea Island—and we went once to the Okefenokee for a guided tour. For hours we prowled the dark water at low speed, bathed in the damp heady southern air that always made me glad when I stepped off the plane into its musky perfume. I could not count all the wading birds that stalked in the Okefenokee shallows like tall aloof models. We glided past cypress and their peculiar pointy knees. Our guide said the knees breathed for the cypress. He pulled up to a small island and waved his hand with a grandiose gesture at the mossy ground. I stepped out of the boat onto the island and the ground moved in an undulating roll. It was a mat of sphagnum moss, and although some people say walking on it is like walking on a water bed, I felt its billowy heave was more like a wave of dizziness before you pass out—a very slow falling sensation although you remain upright.

There are many rich and rare names and swamp words in Meanley's book—*pocosin* and *hammock*, *Choctaw logs*, *gator-hole*. *Pocosin*, says Meanley, is an Indian word that means "swamp-on-a-hill."

Pocosins are almost impenetrable evergreen bog-shrub depressions that some thought were made by meteorites.

Meanley remarked on one pocosin he found while scouting blackbird roosts near Pinetown in Beaufort County. This was likely Big Pocosin, described by B. W. Wells, whose family cleared 50 acres of the gum-maple swamp pocosin "before 1855" and in 1915 another 50-plus acres of "mature swamp forest . . . I saw the trees cut and was at an old-time log rolling where groups of men piled large gum and maple logs into enormous piles for burning."

When Meanley was roaming the swamps there were still 2 million acres of pocosins in North Carolina. By the 1980s only 700,000 acres were left after drainage for forestry and farm. Today government and private projects are restoring unique pocosins such as North Carolina's coastal plains Pocosin Lakes National Wildlife Refuge.

As the scientists put it:

> . . . restoring healthy pocosin wetlands provides important benefits to terrestrial and aquatic ecosystems, as well as human communities: They provide wildlife habitat, lessen the frequency and severity of wildfires, sequester carbon, nitrogen, and mercury (known as a carbon sink), protect the water quality of estuaries, and control flooding in low-elevation coastal areas. Pocosin restoration also plays a key role in the adaption of ecosystems to sea level rise by preventing soil loss and promoting soil formation.

My best near-swamp experience came one summer when I lived in a remote and ramshackle house in Vermont with a beaver-populated swamp half a mile down in the bottom. I went to the swamp almost every day by a circuitous route through the woods, passing a patch of pitcher plants and two or three sundews, across a brook, following beaver tree-drag ruts to an old stick dam. There were trout in this swamp and beautiful painted turtles. I watched the amazing acrobatics of dragonflies with disbelief that they could actually be doing what I saw them do. Even when I sat on the back porch high above the swamp I thought I could catch the green smell of bruised lily pads. I thought constantly about this swamp when I was not in it and still remember it with longing. Once, after weeks away, I came back to the old house in late afternoon. On the plane I had started reading Norman Maclean's novella *A River Runs Through It* for the first time and once at the house decided to read to the end before I went inside. It was an utterly quiet windless golden day, the light softening to peach nectar as I read and ultimately reached the last sentence: "I am haunted by waters." I closed the book and looked toward the swamp. Sitting on the stone wall fifteen feet away was a large bobcat who had been watching me read. When our eyes met the cat slipped into the tall grass like a ribbon of water and I watched the grass quiver as it headed down to the woods, to the stream, to the swamp.

The Dismal Swamp

The country's most famous swamp is known by the morbidly romantic name "Great Dismal." It was a coastal-plain brew of swamp, marsh and open water that straddled Virginia and North Carolina. In the center lay the exquisite freshwater Lake Drummond with its cypress stumps and high-lifted knees looking in twilight like the ruined mansions of some lost civilization. The lake was named for its discoverer, William Drummond, North Carolina's first colonial governor. Legend has it that in earlier years Drummond was one of a hunting party lost in the Great Dismal; all but Drummond perished, and he told that his wandering had taken him to a great glittering lake in the center of the swamp.

The wealthy and lettered Virginian plantation owner William Byrd II (1674–1744) thought the swamp—which he described as "this vast body of dirt and nastiness"—could be drained and planted to hemp. He outlined a drainage scheme and advised that to supply the labor of digging the ditches "let 10 seasoned negroes be purchased, of both sexes, that their breed may supply the loss." By "loss" he meant death and wrote crudely, ". . . those which happen to dye, 'tis probable that their places will be fully supplyd by their children."

Byrd oversaw the first survey of the Virginia–North Carolina state line which ran across the Great Dismal. The survey, though fueled by much wine and rum, was declared accurate when retraced in modern times. Byrd's mordant wit colors his 1728 Pepysian account of the trip in *The History of the Dividing Line Betwixt Virginia and North Carolina*, and there was even more creaky Rabelaisian humor in his bawdy *The Secret History* which only surfaced in 1929.

They ran aground in Currituck Sound, waded in the pluff mud of the marshes. Shoebrush broke a tooth on a biscuit and nearly dallied with a black woman in an isolated cabin, Firebrand and several others . . . drank and cavorted with a "Tallow-faced Wench" one night at Balance's Plantation. "They examined all her hidden Charms, and play'd a great many gay pranks," Steddy reported. "While Firebrand who had the most Curiosity, was ranging over her sweet Person, he pick't off several Scabs as big as Nipples, the Consequence of eating too much Pork."

George Washington was (among other things) a land speculator. In his day the swamp was stuffed with white cedar, cypress and tupelo trees, Spanish moss, festooned with vines, twisters and climbers, dozens of fern species and devil's walking sticks. Washington, with eleven other adventurers, had enough chutzpah to try to drain the Great Dismal Swamp (with slave labor) for agricultural land. In 1763 they set up the Dismal Swamp Company. The War of Independence got in the way and the project languished for some years, then revived in 1785 with the idea of bringing in tongue-twister Deutsch ditch diggers. That didn't happen and the goal changed: instead of farmland, a canal to connect to Albemarle Sound and Chesapeake Bay seemed better; farmland was all very well, but there was a greater prize: wide deep ditches could serve as commercial canals transporting cotton, foodstuffs, luxury items, people, timber, crops, firewood—and shingles. The Dismal Swamp Land Company cut cypress and juniper for shingles and sent them to market on flatboats through their canal. Washington hung in for a while, then, in 1795, the year the company cut more than a million shingles, almost sold his shares to the anti-slavery-politician-planter-founding-father Henry Lee, father of Robert E. Lee. But Henry Lee could not raise the money and Washington's share went to his heirs. Five miles of the ditch were completed in 1805. It exists today as "Washington's Ditch."

Such was the poetic allure of the word *dismal* that the Irish poet Thomas Moore wrote "A Ballad: The Lake of the Dismal Swamp" that featured the legend of a ghostly maiden who paddled the swamp in a white canoe lit with a firefly lamp. In the twentieth century twenty-year-old Robert Frost, his marriage proposal turned down by Elinor White, to whom he had just presented his first slim volume of poems, took a train south and headed into the Great Dismal Swamp with the suicidal intention of never emerging. After some time he met a party of hunters and asked to be led out of the swamp, went on to marry the girl and (famously) write poems.

The modern portrayal of the swamp is Bland Simpson's resonant and affectionate 1990 *The Great Dismal: A Carolinian's Swamp Memoir,* loaded with colorful and crotchety characters, yet Brooke Meanley's forty-eight-page essay "The Great Dismal Swamp" was for many years the insider's guide to the geography of the swamp, its plants and birds from logferns to bears, including the swamp's importance in indigenous species habitat ranges. Meanley found that the "most remarkable sight in the Dismal is the winter blackbird roost in the evergreen shrub-bog along the Carolina-Virginia border . . . with an estimated 10 million birds. . . . one million robins also winter in the same section of the swamp."

For generations of botanists the Great Dismal has been the southern swamp that served as a place marker for the northern growth limit of many warm-climate trees and plants. The American south/north divide was the Great Dismal. Now, as the earth heats, the swamp's identifier role will likely shift northward.

Well into the twentieth century drainage projects and free-for-all timber logging diminished the Dismal. One local man remarked on the fallen giants:

> They cut it with crosscut saws.... Trees so big. I've seen hollow cypress back there where they'd cut the tree, and the first log'd be hollow and they'd just leave it lying there where it fell. I'm six foot tall, and I could walk right in that log and not even bow my head sixteen feet one end to the other.

The deforestation of the swamp continued and by mid-century conservationists were looking for ways to save what was left. Rescue came from a surprising direction.

The Dismal Swamp Land Company founded in 1763 had existed for 136 years, paying out tiny inherited dividends to its many shareholders until it finally was sold in 1899 to William Camp. The Camp Manufacturing Company quickly converted forest to large clear-cuts. Later, after a merger, the business became the Union Camp Company, makers of lumber and paper bags. But high water that hindered tree-cutting was a constant problem and in the 1960s Union Camp diversified its business, moving into building products, plastics and real estate. The company leased out some of its land along the new interstate highways as individual sites (Travelers' Oases), each garnished with motel and restaurant. The leases brought in $15,000 per acre a year compared to the pitiful $5 a year for an acre of trees. In 1973 Union Camp looked at its swamplands and made an easy decision to donate 50,000 heavily treed and wet Virginia acres, without much commercial value, to the Nature Conservancy. The acreage lay in the Great Dismal and included most of Lake Drummond. It was then the largest gift ever made to the Nature Conservancy. After negotiations the Conservancy turned the swamp forestland over to the Interior Department's Fish and Wildlife Service to be designated a wildlife refuge and Union Camp deducted $12 million from its taxable earnings. The gift started a domino effect of swamp donations that brought the total to 106,000 Dismal acres. Then began the flood of new ecological, botanical and hydrological studies which continue today: sediment and water quality assessments of Lake Drummond; models of carbon balance after disturbances of storm, fire, flood, vegetation clearing; peat fire mitigation data after "catastrophic" fires in 2008 and 2011 showed the swamp's vulnerability; studies of carbon and methane fluctuations; DNA-based black bear hair samples to identify population levels and road-crossing places; aerial color and infrared photographs for analysis of vegetation communities; hydrological restoration—and more.

Once gurgling through more than 2,000 square miles, the Dismal today is compressed into 175 square wilderness miles of "The Great Dismal Swamp National Wildlife Refuge." It remains an important place for black bears and the sky above is still the route of the ancient Atlantic flyway for migrating birds.

The Great Black Swamp that covered much of Ohio and parts of Michigan and Indiana inspired deep and lasting hatred. After the Lewis and Clark expedition (1804–1805) and the Erie Canal's gradual opening from 1825 onward, the country's swelling population pushed west into the new territories. Travelers forced to splash through swamps under attack from blackflies, no-see-ums, deer flies and mosquitoes or make long tiresome detours around watery areas complained vociferously and called to the heavens for drainage.

The Great Black Swamp, an excess of mire left over from the southwest portion of glacial melt Lake Erie, was notoriously troublesome at forcing arduous detours. The Black Swamp froze itself blue in winter and simmered under summer sun. It was 40 miles wide and 120 miles long, an elm-ash watery woodland well stocked with snakes, wildcats, moose, insects, birds, malaria-carrying mosquitoes and unnamed demons, immovably in the way of all who were trying to go west.

By the 1850s farmers noticed that the raised stream banks in parts of the swamp were made of dry black soil. They picked up handfuls of it, rubbed it between their fingers, judged its tilth. Then they cut down the stream bank trees, plowed and planted and harvested tremendous crops. They said what every farmer in newly opened peatland has ever said as they gathered the first harvests: "this is some of the most productive soil on earth." Other farmers noticed, and since stream bank acres were limited a few men with experience of wet soils tried drainage with ditches and tiles. Excited by their success the farmers attacked the Black Swamp; a mad make-your-own-land rush was on. In the 1880s an Ohio man, James B. Hill, frustrated by the slow work of laying drainage tiles, invented a machine he called the Buckeye Traction Digger. Farmers bought or borrowed them, and the Black Swamp began to dry up.

Pro-drainage legislation helped the drainage process along and woe betide the landowner who resisted his neighbor's drain work. The legal maneuverings make interesting reading. For example, in 1915 Ben Palmer of Minnesota wrote a legal guide to drainage. Chapter IV—"Drainage Legislation and Adjudication"—opens with these sentences:

> Thirty-six states of the Union have now enacted general drainage laws for the purpose of providing the legal machinery which is necessary if drainage work involving any considerable amount of land is to be successfully carried on. These laws apply to lands which can not be drained or protected from overflow by their owners without building ditches across the lands of others, so that it becomes necessary to provide a system of procedure that will enable the more enterprising proprietors to cooperate in draining their farms without being blocked in their efforts by a small minority who refuse to allow ditches to be built across their lands. These laws also aim to insure adequate drainage outlets, remuneration for property taken or injured for the common good, and an equitable distribution of the costs of the work.

By the early twentieth century only a pinch of the original Black Swamp still existed—the rest was "some of the most productive soil on earth." It was taken as a stroke of luck that the drainage tiles could be made from the clay deposits beneath the rich peaty soil—in a way the Black Swamp paid for its own annihilation. But a few generations later the productive soils were depleted; organic soils disappear when they are not replenished. Manure grew scarce as tractors replaced horses. The farm world welcomed synthetic fertilizer. Time passed, and the Maumee River which drains the Ohio cropland watershed became a major source of pollution in Lake Erie. I was once on a train that stopped for hours on a bridge over the Maumee River to let freight traffic through. There was no sign—frothy scum, iridescent gloss or bright algae—to show that just below the train flowed Lake Erie's poison enemy.

Aside from the joys of draining, there was another pot of gold at the end of the swamp—fortunes for the nineteenth-century woodlands owners and professional timbermen who cut down the wetland forests not only of Ohio, but Michigan, Indiana, Illinois, Georgia, Louisiana, Florida, any state north or south that had swamp forests—mixed forests of irreplaceable giant elm, ash, oak, birch, poplar, maple, basswood, hickory and chestnut. They were just a few decades behind the Great Dismal's lumbermen who cut and split the handsome cypress into roof shingles.

Sharon Levy, a science writer who specializes in current water and wetland issues, wrote feelingly of the mark the Black Swamp made on Ohioans.

The tough people who conquered the Great Black Swamp did so at great personal expense, and they've passed down a deep and abiding loathing of wetlands. They are considered a menace, a threat, a thing to be overcome. These attitudes are enshrined in state law, which makes impossible any action, including wetland restoration, that slows the runoff through those miles of constructed drainage ditches—the very conduits that, after each heavy rainfall, deliver thousands of metric tons of phosphorus and nitrogen to the Maumee, and onward into Lake Erie from which millions of people drink.

Although the water authority William Mitsch has suggested that if 10 percent of the old Black Swamp soils were allowed to become wetlands again they would cleanse the runoff, Ohioans still are powerfully anti-wetland. Even private efforts to restore small wetland areas are met with complaints from neighbors about noisy frogs and fears of flooding. Yet some do care; the Black Swamp Conservancy in 1993 set up a conservation land trust of 19,000 acres kept carefully reined in by the land stewards and farmlands around it.

The Kankakee Marsh

Indiana people mourn the loss of their Grand Kankakee a century after it was obliterated. Of the Kankakee William Mitsch remarked, "drainage was absolute." Northwest Indiana's Kankakee was an extensive swamp-marsh of more than 500,000 acres on a sandy dune outwash plain, in retrospect called "one of the great freshwater wetland ecosystems of the world." The explorer Rene-Robert Cavelier, Sieur de La Salle, had explored the upper basin of the Mississippi River and traveled down the Kankakee in 1679. The great swamp-marsh then was both stunningly beautiful and full of diverse birds and wildlife in almost unbelievable abundance: mink, otter, skunks, ducks, migrating bird flocks, muskrats, bass, walleyes, frogs, turtles, passenger pigeons, minks, bobcats. Later important people came from abroad to this "sportsman's paradise" to shoot and shoot and shoot the endless streams of waterfowl. The wildlife departed for the Chicago markets and restaurants (and taxidermists) in full-packed railroad cars.

The Kankakee River snaked its 250-mile way through the

swamp in two thousand twists and bends, a slow absorbent river punctuated with bayous and edged by riverine forests. In the storied days of the fur trade the strategic St. Joseph portage on the Kankakee was known to every Indian, fur trapper, explorer, Jesuit missionary and voyageur. In a certain stretch the river formed heart-shaped islands, one of which was a favorite of the Kankakee-born naturalist Charles H. Bartlett, whose 1907 *Tales of Kankakee Land* has kept the region's love for the lost swamp-marsh alive.

> This island, whose quiet haunts we loved to invade, was covered . . . with an oak-grove, with here and there a giant shell-bark hickory. The soft turf spread beneath this grove was screened from view on all sides by the tops of dense thickets of dogwood, and marsh maples and soft willows that rose from the low ground surrounding the island, their upper branches glancing over into the higher plain which they could not invade. Here and there, over the interior, was a clump of sassafras or a billowy area of wild roses . . . a place where a few white birches lifted their graceful, though ghostly, forms. . . . [In a] boggy indentation . . . there stood a dark compact mass of tamaracks . . . in exquisite contrast against the dull gray wall of massive oak trunks that leaned from the top of the banks . . .

The United States had acquired the Kankakee through treaties with the Potawatomi, and in 1850 gave the swamp to the state of Indiana for drainage and conversion to farmland. The newly drained farm fields produced terrific harvests of grains, sugar beets and onions, for the rich organic muck was, of course, "some of the most productive soil in the world." The newly drained land was not farmed by local farmers with smallholdings but by men who somehow received very large swathes of the old swamp. Nelson Morris, one of the three big Chicago meat-packers, looked a little like Groucho Marx, sported a cigarette holder and managed to acquire 25,000 acres of drained Kankakee land where he pastured part of his Texas herd. A local man, Lemuel Milk, got 10,000 acres and a nephew cornered another 4,000 acres. This scenario has been repeated the world over: swathes of fen, bog or swamp are deemed too wet for agriculture and the cry goes up that for the public good it must be drained. But the new lands then usually became the property of developers and big agriculturists or ranchers—public good neatly sidestepped.

When much of the Kankakee swamp (aka the "Everglades of the north") was tamed to obedience, attention turned to the river. The Erie Canal and parts of the Great Dismal had been hand-dug, but the Kankakee's destruction fell to the jaws of one of the period's major innovations—the steam dredge.

In 1902 these monstrous dredges started the work of gouging out ruler-straight channels, ignoring the river's two thousand natural bends. Excavated material was dumped along the edges of the ditch to serve as levees or transport paths. Lumbermen then went in and cut the big hardwoods—oaks, walnuts, elms, sycamores—and hauled them out through the handy new canals. When they finished obliterating the Kankakee, the new unbending canal system was ninety miles long, a mere 36 percent of what had been the river's varied and complex natural length of 250 miles. Straightening the ambling river speeded up the water flow and the stream's greater velocity began to erode the banks. The roots of the trees along the somnolent old banks had held the soil in place, but with the trees gone there was nothing to prevent the soil from washing away. The banks flattened out and became lower, easily topped in periods of heavy rain and snowmelt. The erosion and flooding were unpleasant surprises for Kankakee residents and these became modern problems that still eat up large amounts of time and money.

It is, of course, possible to love a swamp. I remember a small and nameless Vermont larch swamp that could be reached only by passage through a gloomy ravine that I thought of as the "Slough of Despond." At the bottom of the ravine ran Jacob's Chopping Brook. The flurried emotional water of the brook contrasted with the black glass disc of swamp water that seemed made to reflect passing clouds but under rain showed itself as dimpled pewter. It has been fifty years since I last saw it, but it is still with me. So I can sympathize with the modern residents of northern Indiana and next-door Illinois over loss of their great river-swamp-marsh. Through the years there have been cries for restoration of the lost landscape and at the end of the twentieth century people were working out ways to do it. It is an important decision to restore even a small piece of wetland that has been severely mauled. Bogs and swamps take thousands of years to build up and develop; humans and their machinery can wipe out those centuries in a few months. Once land is apportioned to owners there can be no easy path to restoration of a natural habitat. The uncomfortable tale of the Everglades is that restoration has dragged on and on for many years in bits and pieces, fighting for money and against loud opposition—as the red queen said in *Alice's Adventures in Wonderland*, "it takes all the running you can do, to keep in the same place." Indiana started making plans for Kankakee restoration in the late twentieth century and nearly thirty years later is living with many complex restoration plans in at least ten counties. The work is supported by local, state and federal governments, dozens of organizations including Ducks Unlimited, the Indiana Heritage Trust, the Nature Conservancy, park departments, forestry organizations and private companies and by ordinary people.

One of the swamp areas restored by the Nature Conservancy annually attracts thousands of sandhill cranes and thousands of human viewers of the mighty clanging throngs. Aldo Leopold wrote of the crane, "When we hear his call we hear no mere bird. We hear the trumpet in the orchestra of evolution. He is the symbol of our untamable past, of that incredible sweep of millennia which underlies and conditions the daily affairs of birds and men."

The Limberlost

The comparatively small (two miles wide and ten miles long) Limberlost Swamp in northeast Indiana several miles from the west side of the Great Black Swamp illustrates a pervasive nineteenth-century mind-set. Even such presumed American "nature" novels as Gene Stratton-Porter's *Girl of the Limberlost*, my mother's favorite book in the 1920s when she was a teenager (she loved it for its swamp setting), is the familiar American story of taking from nature. The 13,000-acre Limberlost Swamp near Porter's Indiana home, though small, was still a diverse and complex system of streams and ponds eventually draining into the Wabash River. The Limberlost was made up of timber, reeds, sphagnum moss, orchids, sundew, pitcher plants and grasses that nurtured great crowds of water- and migratory birds, snakes, frogs and other amphibians, deer, muskrat and beaver, mink and an encyclopedia of insects, including rare moths and butterflies.

There are two and probably more stories of how the name "Limberlost" originated. In one, a man named Jim Miller, so physically agile he was called "Limber Jim," was hunting in the swamp. He became hopelessly lost, walked in deadly circles before he began to blaze trees in a straight line. His friends found him and referred to the swamp ever after as the place where Limber was lost. Another story refers to Limber Jim Corbus (what is it with these flexible Indiana men?), who also set out for a day's hunt in the swamp and became lost, but blazed no trees and was never found.

Girl of the Limberlost championed Elnora, who collected the chrysalides of moths, raised, killed and mounted them. After her first miserable day in high school, where she was scorned as an out-of-fashion backwoods hick, she saw a placard in the local bank window offering cash for moths, cocoons and pupae cases. Elnora needed money to buy nice clothes and cosmetics that would let her join popular high school cliques and pay for her books. She described her moths to the placard writer, the Bird Woman, who told her, "Young woman, that's the rarest moth in America. If you have a hundred of them that's worth a hundred dollars according to my list." Elnora was on her way to wealth, a career and all the rest of it thanks to the corpses of the yellow emperor moth.

Against Porter's protests the Limberlost was ruinously drained for farmland by steam-powered dredges from 1888 to 1910. But in the 1990s Indiana readers who treasured Porter's book bought up some of the original swamp acreage, and with help from several conservation groups started restoring the swamp by removing drainage tiles. As the water deepened they planted native sedges, grasses, trees and water plants. Today a small piece of the Limberlost exists again, a tourist attraction and home to muskrats, ducks, herons, turtles, fish and insects. The yellow emperor moths are still around and not on endangered lists, though declining in number, in part because they are said to be deeply annoyed by streetlamps.

In the nineteenth century setting tile pipes, ditching and draining wetlands was a sign of progress. The great Horicon Marsh in Wisconsin was another irritant to travelers until it was dammed in 1846 and became navigable by boat. It was undammed and returned to natural marsh in 1869. A few years later a sportsmen's club happily estimated that 500,000 ducks hatched out in the marsh every year. It is one of the ironies of wildlife conservation that hunters like Ducks Unlimited have done the most to provide habitat for wild waterfowl.

Mangrove Swamps

Mangroves are marine trees. They grow in brackish and saline water along southern and tropical shores, their splayed-out roots resembling the "cages" that supported Victorian hoop skirts—and they form peat. Their specialized home ground is salty, brackish, smelly and muddy. There are roughly sixty species of mangrove, most in Asia, and the strongest forests are those of mixed species. Mangrove swamps have been called the earth's most important ecosystem because they form a bristling wall that stabilizes land's edge and protects shorelines from hurricanes and erosion, because they are breeding grounds and protective nurseries for thousands of species including juvenile barracuda, tarpon, crabs, shrimp, shellfish. Mitsch succinctly describes the mangrove forest as a place "infamous for its impenetrable maze of woody vegetation, its unconsolidated peat that seems to have no bottom, and its many adaptations to the double stresses of flooding and salinity." The trees take the full brunt of most storms and hurricanes—but not all. Hurricane Irma in 2017 hit the mangroves of Big Pine Key in Florida. Trees and shrubs came back after a time but the mangroves did not, presumably because the storm surge plastered a very fine coating of sediment on the vital aerial roots which dried into a choking hard sealant.

Mangrove leaves fall into the water and as they decay become the base for a complex food web benefiting algae, invertebrates and the creatures who feed on them, such as jellyfish, anemones, various worms and sponges and birds. The peat that mangroves form is especially soft and deep, ideal for clams and snails, crabs and shrimp. The mangrove's roots filter out harmful nitrate and phosphate pollutants. The tangled branches above the water make a safe habitat for literally thousands of species, including insects which attract birds. They offer resting places to migrating birds, nesting places for kingfishers, herons and egrets. Monitor lizards, macaque monkeys and fisher cats on the hunt prowl the branches. Below the water the knots of interlaced roots protect tiny fish from ravenous jaws of larger fish, and even manatees and dolphins take refuge there. Mangroves interact with coral by trapping muddy sediments that would smother the reef while the offshore reef protects the mangroves and seagrass beds from pummeling waves. Structurally mangroves form an enormous hedge that extends down into the water and high above it. They are a major part of the "blue carbon" group that absorbs CO_2—the salt marshes and sea grasses, kelp and other seaweed beds.

With all of these virtues it would seem that mangroves must be the most valued trees on earth. Unfortunately that is not so. Although climate researchers see mangrove swamps as crucially important front-line defenses against rising seawater and as superior absorbers of CO_2, five times more efficient than tropical forests, despite their importance and benefits mangrove swamps are in big trouble. Their enemies include industrial shrimp farms, developers hungry to root them out of prime real estate locations, and in Mexico, a country with extensive mangrove forests, they are being deliberately destroyed at the behest of President López Obrador to open an area for the construction of a large Pemex oil refinery. Cynics (I am one) note that Mexico was one of the signers of the Paris climate agreement.

In 2010 a count showed about 53,000 square miles of mangrove forest protected earth's coasts. But six years later another 1,300 square miles of mangroves had been lost to palm oil and rice farms and shrimp aquaculture. In some cases mangrove forests have been removed to make room for shrimp ponds, in other places the shrimp ponds are set back from the mangroves, but the released effluents and pollution still damage and degrade the mangrove forest by changing the water salinity, altering the mangrove's ability to take in nutrients. The result is slow death for the mangroves.

Indonesia has a bad mangrove-loss problem even greater than the vast deforestation of its tropical woodlands. As the mangroves die the local fish numbers crash, the fishery waters become emptier, more coastal soil erodes, more CO_2 and methane are not absorbed. Indonesia is painfully aware of its problems. Through the Indonesian Peatland and Mangrove Restoration Agency the country is struggling to restore its degraded coastal mangroves and coral reefs with help from the World Resources Institute and the World Bank.

Many countries have tried to master the complexities of mangrove restoration, with mixed results. Choice of the right site and a mutually beneficial mix of species is critical. Some well-intentioned restorers planted greenhouse-raised single-species saplings in mudflats where mangroves had never grown, or which were exposed to erosion and strong waves. Mudflats have low oxygen supply because they are constantly wet, and mangroves need to breathe. An expensive World Bank–funded project in the Philippines in the 1980s made most of these mistakes with a survival rate fifteen years later of less than 20 percent of the trees.

The most successful mangrove restoration approach was that of the Florida biologist, ichthyologist and wetlands ecologist Roy "Robin" Lewis III (1952–2018), who worked out the vital details of mangrove increase. Repetitive observation can unravel the mysteries of events and processes. Lewis was still a graduate student when he began working in mangrove swamps. "I spent a decade working in the mangroves before I started to have an understanding of what was going on." He spent years puzzling out the rhythms of mangrove replication. He had observed that in the natural order when a mangrove tree died, plentiful seeds from nearby healthy mangroves floated in and rooted themselves. The problem was location. Just any random part of a shoreline would not work. The flow of water had to be correct. Mangroves need to be sometimes wet and sometimes dry. Lewis worked out a wet-dry ratio of 30:70. Mangroves like wet roots 30 percent of the day and dry roots the rest of the time. "They have a short period of wetness, and then they have a long extended period of dryness, and those alternate daily. . . . That's the secret: you've got to replicate that hydrology." And it was important to involve the local community familiar with mangrove habits and needs.

The first trial of his theory came in 1986 with 1,300 acres of damaged and dead mangroves half-smothered in dirt and weeds near Fort Lauderdale. The flat site had poor hydrology for mangroves, with large areas of standing water. Working with a hydrologist Lewis studied the water situation. After several years of experiment he brought in earthmoving equipment to create a gentle land slope that would allow the natural tide waters to ebb and flow. Then he waited. The tides brought mangrove seeds which took root and five years later three local species of mangroves were growing, fish moved into the sheltering roots and the birds followed. No mangrove saplings were hand-planted; all the new trees grew from water-borne mangrove seeds.

Lewis's way of working with nature—observation and study, planning and patient waiting—has become the gold standard for mangrove restoration, yet many organizations and governments in hope and expectation still plant nursery-started seedlings in the wrong places—and watch them fail.

Other Swamplands

As a species humans in many countries have made mistakes with the swamp wetlands—they were far more sensitive to damage than we knew. The world needs the great swamps we have drained away and the few that still exist but the human impetus to develop and drain continues. Yet until recent decades we didn't even recognize the big tropical peat basins of Amazonia, Indonesia, Peru and the Congo. Attention was focused on the northern peatlands, and ecologists still do not completely understand the workings of those boreal swamps. Current research does focus on them with considerable urgency as the permafrost slumping speeds up.

Some countries promote the idea of the Swamp Beautiful. Once a fabulous wetland like the Pantanal becomes a Ramsar or a World Heritage choice where tourists are guaranteed to see jaguars and other rare animals the swamp almost automatically goes on the lists of the elite tourism companies specializing in exotic locales. Luxury accommodations, guaranteed photo ops, houseboats, safaris, mosquito nets, guided tours and floods of moneyed tourists are part of the rich package that comes with the coveted designation. Nature tamed for tourism seems a variation on museum displays.

The tropical swamp forests hold as much as a third of the earth's underground carbon reserves. Cutting, burning and converting these forests to palm nut oil plantations, as people are now doing in Indonesia, releases gargantuan amounts of CO_2. Although the peatlands of Scandinavia and northern Europe, Alaska and Russia also hold high concentrations of CO_2 and methane gas, in recent summers of the last decade they suffered great fires in Alaska, Greenland and Siberia.

Aside from the release of huge amounts of CO_2 the wind carried black carbon particles from the fires to glaciers, reducing their sunlight reflectivity.

The world's largest wetlands such as Wasur National Park in Papua, New Guinea, the Pantanal in South America, the Kakadu Wetlands in Australia, the Vasyugan Mire in western Siberia, the massive Congo Basin forests, the Kerala backwaters on the Malabar coast of India, the Florida Everglades and the Okavango Delta in Botswana are almost always a mix of fen, bog and swamp—the inclusive word for such mixed peatlands is *mire*. Some of the large wetlands are healthy, most are compromised or damaged. There is a squabble over which swamp is larger, the Gran Pantanal or the Great Vasyugan Mire.

The Gran Pantanal is a river basin in the center of South America, roughly 80,000 square miles of rivers, swamp, rain and drought. The great wetland alternates (as does the Everglades) between flood and the dry season, when it becomes more of a savanna. The Pantanal has been called the richest place on earth in birds, with 463 species recorded, including its iconic jabiru, a giant stork almost five feet tall with an eleven-inch beak. There are caiman, capybara, jaguars—and tourists. A small section of the Brazilian portion is listed as a UNESCO Heritage site, but the Pantanal is ringed by human activities as the forests fall to poachers, agricultural and cattle ranches move closer, gold and diamond miners blast and dig. In addition to mining there is pollution, cocaine smuggling, people capturing rare birds and animals for the illegal wildlife trade. Enforcement of laws is difficult and expensive in this precious and still-remote swamp. Will the Pantanal also suffer the fate of the Amazon—to become a carbon-emitting accelerator of climate change?

The western Siberian Vasyugan, said to have formed ten thousand years ago, is the giant wetland of the northern hemisphere. In tandem with the Pantanal it was considered a major force in adjusting the earth's atmosphere. This massive (and still growing) 370-mile-long swamp lies between the Ob and Irtysh rivers with innumerable side branches enlarging its width to 280 miles. The Vasyugan River feeds the mire and gives it a name. It was nominated for a place on the World Heritage list in 2007 and the application stressed its "extreme complexity" and special types of mire formations and peat deposits, yet a deterrent for acceptance to that important list might be the oil and gas extraction industry in the western part of the swamp. Bird migrations have followed the Vasyugan's sinuous curve for thousands of years and it provides unparalleled habitats for rare and endangered plants and animals. There have been sightings of the globally threatened aquatic warbler and the only known nesting site for the slender-billed curlew, teetering on the brink of extinction, was recorded in the Vasyugan in the years 1909–1925. The bird itself was last seen in Morocco in 1995. A video—the only one—of these rare birds exists.

Originally the Vasyugan was nineteen different wetlands that merged in the Holocene, and its poorly understood water patterns are now of great interest. The old studies focused on ways to exploit its swamps and bogs but according to the biologist Sergey Kirpotkin the new attitude is:

To better understand how the mire originated and developed, how the mire patterns have self-organized, and how the macro landscape has reacted on and counterreacted climate change. Of special importance is the question to what extent the mire will still further fulfill its possible role . . . as a resilient cooler and "border control" against climate change, or whether it better should be treated as a carbon bomb, waiting to be ignited.

It is easy to think of the vast wetland losses as a tragedy and to believe with hopeless conviction that the past cannot be retrieved—tragic and part of our climate crisis anguish. But as we see how valuable wetlands can soften the shocks of change, and how eagerly nature responds to concerned care, the public is beginning to regard the natural world in a different way. The "rights of nature" is a legal concept that is gaining international standing. The United States is among dozens of countries that have committed to some "rights of nature" laws that allow citizens to sue on behalf of lakes, streams, ocean reefs, swamps. The movement started in 2020 with the United Nations inaugural biodiversity summit and is rapidly growing. Of current interest is a first test of the rights-of-nature law passed in Orange County, Florida, in 2020. In April 2021 a consortium of lakes, streams and marshes filed suit in the Ninth Judicial Circuit Court of Florida to halt a planned housing development that would eradicate wetlands and pollute streams. Because of its past exploitative history Florida has become very sensitive to wetland damage. The plaintiffs are Wilde Cypress Branch, Boggy Branch, Crosby Island Marsh and several lakes. It causes a powerful mental shift when you read or say those names as "plaintiffs."

Buying Time

I used to think that stasis in the "natural" world was possible and desirable, but I have learned beliefs like the "balance of nature" are point-in-time-defined fantasies.

We are starting to understand the importance of saving and restoring the peatlands, but there are also catalogues of different waterscapes beyond the scope of these essays that deserve attention: potholes, bottomlands, delta wetlands and gulf plains. The cross-border prairie pothole country was once a bountiful scatter of shallow marshy ponds—waterfowl nurseries—across the Dakotas, Minnesota, Wisconsin, Manitoba, Saskatchewan and Alberta, considered "one of the most important wetland regions in the world because of its numerous shallow lakes and marshes, its rich soils, and its warm summers, which are optimum for waterfowl." The great migration flights of birds depended on the potholes as we depend on motels and diners on long journeys. Missouri and Arkansas had bottomlands on the alluvial plain of the Mississippi. Bottomlands flanked streams and rivers that seasonally flooded and enriched the soil in the same way as the Nile. Mississippi and Louisiana had delta wetland, a complex mixture of river, wetland and uplands, and Texas had its soggy gulf plains where nearly level land was seamed with creeks and rivers. New Orleans was built on swampland but so was Chicago which today is beginning to suffer from floods and rising water as the swamp indicates its eagerness to return and retake its natural place. The isolated, shrunken and maimed Burns Bog near Vancouver in British Columbia is in multiple opposing transitions and subject to continuing drainage even as spotty restoration projects continue.

Louisiana was built up over thousands of years by the sediments the Mississippi disgorged as it spread into the great delta at the Gulf end, a fabulous mixture of salt water and fresh water, mud, marshland, sand, bayous—and in modern times chemical factories that gave the region between New Orleans and Baton Rouge the richly deserved name Cancer Alley. The great Mississippi was famously unruly, flipping out of one channel into another, seeking a comfortable position like a restless sleeper. In 1882 and 1927 catastrophic floods urged the 1928 Flood Control Act through Congress. The entire Mississippi River system was then so heavily reworked with dams, locks and flood control reservoirs that it became a large mud canal. Mark Twain would not recognize this supine ditch. And as melting ice raised sea level, without the deposits of sediments Louisiana began to disappear. The protective marshes that once cushioned New Orleans from storms were underwater. Eighty years after the river was "tamed" Louisiana had lost 1,850 square miles of its shoreline.

In the 1970s scientists presented plans to let the mighty river disgorge some of its mud where it would do the most good. The advice was not taken. In 2005 Hurricane Katrina showed how bad the future was going to be and the fifty-year coastal rescue plan to stave off an inundation that would drown New Orleans went into effect, putting Louisiana in the avant-garde of coping with shifting climate. In 2023 the state will carve out an opening in a levee near Barataria Bay that has held in the river for decades. Once again sediment will spread out, building marshes. It will not be a cure because sea level is irrevocably rising and already the scientists have had to revise their estimate of recovered land downward—the state will gain new coastland but at the same time it will lose more. The coastal geoscientist Torbjörn Törnqvist believes the project is buying time as the climate crisis intensifies: ". . . the difference in the long run between managed retreat and complete chaos."

In the end all humans will be "haunted by waters."

Ω

Acknowledgments

Covid and travel limitations kept me from interviewing scholars and authorities concerned with the peat-producing wetlands. I mostly relied on my personal library, sought-out books and articles to satisfy my interest in fens, bogs and swamps. I learned a great deal from discussions with observant friends, some directly involved with the sciences that examine the natural world. I became interested in underwater archaeology and the preservation of the past in cold waters (and bogs) years ago when I met historian Selma Barkham in Newfoundland. In the Pacific Northwest I benefitted from exciting ideas and knowledge from geologist Katherine Reed, marine science project manager Betsy Carlson, artist bee-fancier Karen Rudd and naturalist Steve Grace. In a discussion of the liminal qualities of ancient bogs with archaeologist Dudley Gardner I heard something of the resurgence of Mongolian shamanism. And in the brief two years of working on these essays climate change picked up speed so rapidly it was not easy to keep up with the torrent of new discoveries. Among newspapers *The Guardian* remained focused on the climate crisis posting almost weekly changes in distant forests and seas. Even the *Siberian Times* ran articles on permafrost slump, the huge Batagai chasm and emerging archaeological discoveries. I ran for the mailbox for the latest issues of *Nature* and *Science* and read dozens of on-line periodicals and newsletters including the brilliant *Hakai Maga-*

zine out of British Columbia and those of the Greifswald Mire Centre. Certainly the research involved in learning about these wetlands was a kind of private enlightenment for me. I especially want to thank my agent, Liz Darhansoff, who thought my private musings might make a book and editor Nan Graham and crew who tackled the messy and awkward pages.

I hope readers gain some of the flexibility of mind that can come with facing up to a world quivering with upheavals of fire, flood, atmospheric rivers, ocean acidity, heat domes and indecisive ocean currents amid the "whirling wheels of change." It is good to remember that this old earth has been constantly changing since it was a spinning mass of outgassing magma.

Notes

1. DISCURSIVE THOUGHTS ON WETLANDS

1 **Wetland** The English word *wetland* did not exist until the 1950s and '60s, coming out of American revisions of hunting laws and bird migrations. *OED* cites *Science News Letter* 1955, *New Scientist* 1965, *Nature* 1969.

4 **psychozoic** Ellen Meloy, *The Last Cheater's Waltz,* (Henry Holt, 1999) p. 69.

4 **centerline** This zigzag central line or stabilimentum is thought by some to reinforce the web's strength against the catastrophic blunder of a flying bird. Interestingly a few years ago I saw at the bird banding station of the Pittsburgh Museum of Natural History experimental samples of window glass made to deter bird collisions. An almost-invisible pattern not unlike a stabilimentum was embedded or etched in the glass.

5 **Childhood image** Bruno Schulz, "An Essay for S. I. Witkiewicz," in *Polish Writers on Writing*, ed. Adam Zagajewski (Trinity University Press, San Antonio, 2007), p. 31. In adult years I read Bruno Schulz's exuberant and surreal stories in *The Street of Crocodiles*. "Read" is not correct: one doesn't read Schulz but swims through his phosphorescent prose.

5 **Schulz's murder** *Ibid.*, p. 31. He was shot by an SS soldier as payback to another SS man.

6 **Place-names** Keith H. Basso, *Wisdom Sits in Places, Landscape and Language Among the Western Apache* (University of New Mexico Press, 1996).

7 **going home** Frank O'Connor, "The Long Road to Ummera," *Collected Stories* (Vintage, 1981).

7 **"swine flu" unfair** Rob Wallace in *Big Farms Make Big Flu* (Monthly Review Press, NY, 2016), p. 34, dislikes the moniker "swine flu" because, he writes, "pigs have very little to do with how influenza emerges. They didn't organize themselves into cities of thousands of immune-compromised pigs. They didn't artificially select out the

genetic variation that could have helped reduce the transmission rates at which the most virulent influenza strains spread. They weren't organized into livestock ghettos alongside thousands of industrial poultry. They don't ship themselves thousands of miles by truck, train or air. Pigs do not naturally fly."

8   **pangolin** R. Frutos, J. Serra-Cobo, T. Chen and C. A. Devaux, "COVID-19: Time to Exonerate the Pangolin from the Transmission of SARS-CoV-2 to Humans," *Infection, Genetics and Evolution* 2020; 84:104493. doi:10.1016/j.meegid.2020.104493.

9   **"the undying difference"** Patrick Kavanagh, "Father Mat."

9   **looking carefully** Years ago in Wyoming a friend and I planted a dozen shrubs at the edge of the cheatgrass-infested patch that served as "lawn." In the early evening a robin flew to the easternmost shrub and walked slowly all around it. The bird went on to the second shrub and repeated the survey and continued down the line until it reached the last plant. That robin likely had personal and intimate knowledge of every plant in the area from sagebrush to the cottonwoods on the river bank.

9   **Thoreau's climate data** Richard B. Primack, *Walden Warming: Climate Change Comes to Thoreau's Woods* (University of Chicago Press, 2014).

10   **alteration in flowering times** *Ibid.*, pp. 45-46. For California flora and fauna scientists similarly use the data-rich field notes of the biologist and zoologist Joseph Grinnell (1877–1939), whose system of recording observations became the gold standard for precise fieldwork. As the earth continues to warm, Grinnell's detailed observations are a baseline to measure the changes.

11   **managing earth** news.mongabay.com/2018/09/.

11   **"big tech"** Naomi Klein, "How Big Tech Plans to Profit from the Pandemic," *Guardian Weekly*, May 29, 2020, pp. 34–39.

11   **root bridges** Julia Watson, *LO-TEK Design by Radical Indigenism* (Taschen, 2019), pp. 46-63.

13   ***writing a novel*** *Barkskins*, 2016. A decade later those forests were stoking the huge conflagrations of 2020.

14   **nature** Genesis 9:2–3.

14   **wetlands always die** Oliver Rackham, *The Illustrated History of the Countryside* (Phoenix Illustrated, 1997), p. 184.

15   **frozen underpinnings** The Svalbard Global Seed Vault, started in 2008, is the Norwegian repository for seeds from humankind's food plants. Deep inside the frozen permafrost the seeds were considered infallibly protected from human wars and disasters. In 2017 climate

crisis warming caused meltwater which leaked into the entrance tunnel. Water is infinitely flexible. Drainage channels and pumps were put in place to forestall future melt-water flooding. But the problem remains.

15 **methane escape** www.theguardian.com/environment/2019/nov/27 /climate-emergency-world-may-have-crossed-tipping-points; www .theguardian.com/science/2020/oct/27/sleeping-giant-arctic -methane-deposits-starting-to-release-scientists-find.

16 **Batagai Slump** *Siberian Times*, March 4, 2021.

16 **Yakutia fires** Anton Troianovski, "As Frozen Land Burns, Siberia Trembles," www.nytimes.com/2021/07/17/world/europe/siberia -fires.html?action=click&module=Spotlight&pgtype=Homepage

16 **end of water** www.semanticscholar.org/paper/The-embarrassment -of-riches%3A-agricultural-food-high-Jefferies-Rockwell/5f18215e 689941d70a2a5c29033992a7a6bd8cf9. Catrin Einhorn et al., "The World's Largest Tropical Wetland Has Become an Inferno," *New York Times*, October 13, 2020.

16 **Ireland** Emily Toner, www.sciencemag.org/news/2018/12/power -peat-more-polluting-coal-its-way-out-ireland

17 **James Rebanks** www.euractiv.com/section/agriculture-food/news/ new-uk-farming-bill-guarantees-subsidies-for-2020. James Rebanks, *Pastoral Song: A Farmer's Journey* (Custom House, 2020).

18 **disappearing earth** Anton Troianovski and Chris Mooney, "2° C: Beyond the Limit, Radical Warming in Siberia Leaves Millions on Unstable Ground," *Washington Post*, October 3, 2019.

19 **fen, bog and swamp definitions** *Fen.* Peatland receiving [mostly ground] water rich in dissolved minerals; vegetation cover composed dominantly of graminoid species and brown mosses.
Bog. (*Muskeg* is the word used most in Alaska and Canada.) Peatland receiving water exclusively from precipitation and not influenced by ground water; sphagnum-dominated vegetation.
Swamp. Peatland dominated by trees, shrubs, and forbs; waters rich in dissolved minerals.

19 **reference books** The basic textbook in the U.S. is Mitsch and Gosselink's *Wetlands*, 5th ed.

19 **Alaska** M. Gracz and P. H. Glaser, *Wetlands Ecology and Management* (2017) 25: 87. doi.org/10.1007/s11273-016-9504-0.

20 **Bog beauty** For those who doubt the beauty, the photographs illustrating Shanna Baker's September 19, 2016, article in *Hakai*, "The Secret World of Bog" in the on-line publication, are evidence.

20 **ecological details** The CWCS sorts the peatlands out as bogs, fens and swamps.

21 **overlooked carbon sink** Stephanie Wood, "Blue Carbon: The Climate Change Solution You've Probably Never Heard Of," *The Narwhal*, September 30, 2020.

21 **bigger than the forests** Now instead of absorbing it the damaged Amazon emits CO_2 Nor did I recognize the importance of peatlands in the years I spent working in *Barkskins*. Now the pendulum has swung back and the idea of planting six billion trees to absorb CO_2 is gaining strength although it will far from solve the multiple problems of climate change.

21 **Amazon basin** Alan Graham, *A Natural History of the New World* (University of Chicago Press, 2011), pp. 93–94.

21 **diversity** Marco Lambertini, *A Naturalist's Guide to the Tropics* (University of Chicago Press, 2000), p. 41.

22 **other ecological parts** Yet another facet of the Amazon complex is the Pantanal, a seasonal flood plain that is the largest wetland on earth, a kind of wet Serengeti in respect to diversity of wildlife. In it is the Victoria waterlily with pads as large as a double bed, jaguars, giant river otters, capybara, brilliant blue macaws, caiman, anaconda. In recent years the Pantanal has become a tourist attraction with all the usual development and crowding that such a fate brings. In 2020 extreme and unprecedented fires have burned at least 10 percent of the Pantanal. Many of its rare and threatened animals have died.

22 **complexity of Amazon** Graham, *op. cit.*, 93–94.

23 **at risk** Dom Phillips, "'Project of Death': Alarm at Bolsonaro's Plan for Amazon-Spanning Bridge," *Guardian*, March 10, 2020. Renata Ruaro et al., "Brazilian National Parks at Risk," *Science* 367, no. 28 (February 2020), p. 990. See also Renata Ruaro et al., "Brazil's Doomed Environmental Licensing," *Science* 372, no. 6546 (June 2021), p. 1049.

23 **teetering on the edge** Fiona Harvey, "Amazon Near Tipping Point of Switching from Rainforest to Savannah—Study," *Guardian*, www .theguardian.com/environment/2020/oct/05/amazon-near -tipping-point-of-switching-from-rainforest-to-savannah-study.

24 **Amazon flip** Luciana V. Gatti et al., "Amazonia as a Carbon Source Linked to Deforestation and Climate Change," *Nature* 595 (July 15, 2021).

24 **Conquistadores** Marie Arana, *Silver, Sword & Stone* (Simon & Schuster, 2019), p. 73 ff.

24 **the Extremadura** *Ibid*.

24 **invaders** John Hemming, *The Search for El Dorado*, cited in Arana, *Silver, Sword & Stone*, p. 50.

24 **beauty meant nothing** Lopez's account is given in Graeme Gibson's *The Fireside Book of Birds* (Doubleday, 2005), pp. 261–63.

25 **Ignorance of strangers** John Hemming, *Tree of Rivers: The Story of the Amazon* (Thames & Hudson, 2008), p. 21. Steven Mithen in his 2003 *After the Ice*, p. 354, mentions two anthropologists who lived with a tropical jungle community in the 1970s. Gathering food in such a forest would be difficult without the habitat-acquired skills: "They hunted monkeys and birds with blowpipes; they caught turtles, tortoises, frogs, fish, prawns and crabs; they dug wild tubers from the forest floor and collected an immense array of ferns, shoots, berries, fruits and seeds."

26 **gold fever** Alvaro Mutis, *The Adventures of Maqroll* (Harper-Collins, 1990), p. 45.

26 **Orellana's end** Hemming, *Tree of Rivers*, pp. 27–34.

26 **Diversity of indigenous people and their cities** Hemming, *Tree of Rivers,* p. 17. Recent evidence for a different kind of urban entity in the tropics, quite unlike the dense modern cities we know, is presented in a new study: Patrick Roberts's *Jungle*, Basic Books, 2021, 153-171. Roberts describes huge spread-out low-density cities of small farms, forest gardens, managed water reservoirs, flood control and soil amendment that existed before European explorers arrived. In the Amazon alone 8 to 20 million people lived this way. Other tropic jungles supported low density cities Angkor Wat, Tikal and Calakmul in the Classic Maya, Anuradhapura in Sri Lanka.

27 **stunned in Amazonia** Henry Walter Bates, *The Naturalist on the River Amazons*, p. 15. "Sipo" is from the Portuguese cipó, for liana.

29 **lightning or arson?** Rachael Kennedy et al, "'Low Chance' Siberia Wildfires Will Be Brought Under Control: Greenpeace Fire Expert," euronews, www.euronews.com/2019/08/06/.

29 **the Pantanal on fire** Catrin Einhorn, "The World's Largest Tropical Wetland Has Become an Inferno," nyt.ms/34LVA7a.

30 **early account of zombie fires** Kate Marsden, *On Sledge and Horseback to Outcast Siberian Lepers*, 1891 (Cambridge University Press reprint, 2012), pp. 103, 136–37.

31 **how much carbon?** W. Hubau, S. L. Lewis, O. L. Phillips et al., "Asynchronous Carbon Sink Saturation in African and Amazonian Tropical Forests," *Nature* 579 (2020), pp. 80–87. As a nonacademic

I am denied access to the full article. However, many unsequestered publications reported on the content.

33 **"Blue Babe"** R. Dale Guthrie, *Frozen Fauna of the Mammoth Steppe, the Story of Blue Babe*, 1990. The chapter "Unearthing Blue Babe" is a gripping step by step account of the meticulous examination of the physical evidence that led to Guthrie's surprised conclusion that Blue Babe was killed by Alaskan lions 36,000 years ago.

33 **Vermeer** www.essentialvermeer.com/palette/rare.html.

33 **"imperceptible"** I follow the French historian Fernand Braudel's inclusive perspective on puzzling out history through *la longue durée*: www.oxfordreference.com, *longue durée*.

34 **sandy bluff** The bluff here referred to lies west of Wilson's Point lighthouse at Fort Worden State Park, Port Townsend, Washington. I thank the geologist Kitty Reed for explaining the intricacies of the bluff. There is a useful guide—"Geology of the North Beach Bluff, Fort Worden," issued in 2017 by members of the Quimper Geological Society.

2. THE ENGLISH FENS

35 **fen definition** William J. Mitsch and James G. Gosselink, *Wetlands*, 5th ed. (Wiley), p. 711.

38 **North and South fens** North Lincolnshire and Yorkshire were the Northern Fen, south Lincolnshire and Cambridgeshire the Southern Fen.

38 **importance of fenlands** Eric H. Ash, *The Draining of the Fens, Projectors, Popular politics, and State Building in Early Modern England* (Johns Hopkins University Press, 2017).

39 **The Acerbic Hand** Reprint of Samuel H. Miller and B. J. Skertchly, *The Fenland Past and Present* (London, 1878), pp. 122–23.

39 **Collector's madness** Christine Cheater, "Collectors of Nature's Curiosities: Science, Popular Culture and the Rise of Natural History Museums," in *Frankenstein's Science: Experimentation and Discovery in Romantic Culture, 1780-1830*, ed. Christa Knellwolf and Jane Goodall (Ashgate, 2008), p. 167.

40 **repetitive robbery** Bernd Brunner, *Birdmania*, 2015, p. 190.

40 **the history of everything** Cheater, "Collectors," in Knellwolf and Goodall, *op. cit.*, p. 168.

40 **early beetlemania** Michael A. Salmon, *The Aurelian Legacy: British Butterflies and Their Collectors* (University of California Press, 2000), p. 26.

40 **excitement of the chase** *Ibid*, p. 27.

41 **Nabokov as lepidopterist** Vladimir Nabokov, *Speak, Memory*, 1966 rev. ed., p. 126.

42 **Paradise lost**, *Ibid.*, pp. 17–18.

42 **isolate refugia** Ian D. Rotherham, *The Lost Fens: England's Greatest Ecological Disaster* (The History Press, 2013), p. 85.

42 **values** *Ibid.*, p 285.

43 **Fen people worn down** Ash, *op. cit.*, *passim*.

43 **progress** C. Robert Haywood, *Trails South: The Wagon Road Economy* (Norman, OK, 1886, 2006), p. 13.

44 **Finn Mac Cool** Flann O'Brien, *At Swim-Two-Birds* (New American Library, 1951, 1966), p. 17.

44 **Doggerland** The awkward name for the inundated territory—Doggerland—was a reference to a rich fishing ground—Dogger Bank—itself named after the two-masted dogger boats of Dutch fishermen who frequented the shoal waters off eastern England's coast. For the museum exhibit: www.youtube.com/watch?v=a3PzgSJT1bU.

45 **they adapted** Luc Amkreutz, quoted in Andrew Curry's "Europe's Lost Frontier," *Science* 367, no. 6477, p. 503.

45 **Storegga Slide** J. Walker, V. Gaffney, M. Muru, A. Fraser, M. Bates and R. Bates, "A Great Wave: The Storegga Tsunami and the End of Doggerland?" *Antiquity* 94 (378), 1409–25.

46 **bog oak** www.theguardian.com/artanddesign/gallery/2019/aug/19/string-theory-make-acoustic-guitar-in-pictures. The English luthier Rosie Heydenrych remarks, "I often use . . . ancient Fenland black oak, which is 5,000-year-old oak and has naturally taken on the dark colouring from the carbon in the bog."

46 **moorlog** "Moorlog" is a Friesan fisherman's word for lumps of peat they hauled up from below. It is interesting that the esteemed Dutch writer W. F. Hermans named an aging student character "Moorlag" in *The Darkroom of Damocles*, 1958, eng. trans. 2007.

46 **Water preservation** Equally astonishing was the preservation of items aboard the *San Juan*, a Spanish ship that sank in Red Bay off the coast of Labrador in 1565.

47 **prehistoric "harpoon"** Vincent Gaffney, Simon Fitch and David Smith, *Europe's Lost World: The Rediscovery of Doggerland,* Research Report No. 160 (Council for British Archaeology, 2009), p. 14.

47 **frontlets** Nicky Milner, Barry Taylor, Chantal Conneller and Tim Schadia-Hall, *Star Carr: Life in Britain After the Ice Age*, Council for British Archaeology, 2013.

48 **shock of recognition** Foreword in Vincent Gaffney and Simon Fitch, *Mapping Doggerland: The Mesolithic Landscapes of the Southern North Sea* (English Heritage, 2007), p. vii.

49 **rapid SLR** Fiona Gruber, "Mammoths and Stone-Age Humans Once Roamed Doggerland, the Lost Land Submerged by the North Sea," www.abc.net.au/news/science/2019-11-20/.

50 **"getting the most"** Paul Shepard, "Nature and Madness," in *Ecopsychology* (Sierra Club Books, 1995), p. 22, quoting Hervey Cleckley's *The Masks of Sanity* (St. Louis, Mosby, 1976), n.p.

50 **"plashy fen"** William Boot's phrase, later lampooned by Evelyn Waugh. For the richest list of British nature writing see Robert Macfarlane's 2005 essay "Where the Wild Things Were."

50 **1 percent remains** *Brittanica*; www.lincsfenlands.org.uk/admin /resources/fens-for-the-future-leaflet.pdf.

51 **development** www.theguardian.com/books/2005/jul/30/features reviews.guardianreview22.

51 **Cornwall hedges** James Rebanks, *Pastoral Song: A Farmer's Journey* (Custom House, 2020). Robert Macfarlane, *Landmarks* (Hamish Hamilton, 2015). Another lost wildness was in the wide historic hedges of Cornwall, some reported to be four thousand years old. These hedges are an interwoven tangle of stone, earth, shrub, tree, flora, birds and small animals, beautiful and representative of the region's remnant wild character. Rebanks makes much of restoring neglected hedges in places housing developers were bulldozing them out at a great rate. I watched hours of the television drama *Poldark* set in Cornwall for the few fleeting scenes showing those hedges. But hedge preservation groups have risen up like reeds, and in 2008 the ancient and secretive Guild of Cornish Hedgers, perhaps chided for keeping what they knew to themselves, created the first manual of hedge construction and repair, guaranteeing that the skills will not be lost utterly.

51 **Brexit influence?** J. R. Ravensdale, *Liable to Floods, Village Landscape on the Edge of the Fens AD 450–1850* (Cambridge University Press, 1974), pp. 196–98. One wonders if Britons, aware of the biennial WWF lists and *Landmarks*, suffered a great sense of sadness and loss. We cannot know if a longing for an irretrievable time of pollarding, eeling, cutting reeds and grasses for thatch, hearing the low-pitched *ka-thump ka-thump* of the bittern influenced the Brexit vote.

51 **landmark words** Word lovers will find more endangered landmark words in the *Dictionary of Newfoundland English*; such specialties as *fish chop* (a cod tail slapping the water), *chuckley* (chokecherry), *horn* (octopus tentacle), *green lick* (a new-cut spruce with running sap).

52 **natural wealth** David Hall and John Coles, *Fenland Survey*, English Heritage Archaeological Report I, 1994, p. 2.

53 **Sweet Trackway** Robert Van de Noort and Aidan O'Sullivan, *Rethinking Wetland Archaeology* (Gerald Duckworth & Co., 2006), 15.

54 **low sea level** Ash, *The Draining of the Fens*, p. 20. One of the *Time Team* BBC television archaeology episodes details the digging of a Mesolithic site threatened by sea level changes.

54 **terrones** Similarly New Mexico people first used *terrones* (slabs of peat cut from sedge meadows near the Rio Grande) and later made hand-shaped and dried adobe bricks.

55 **land ownership** Where did the landless rural populations go? If not to the cities to become part of the urban proletariat, then to Canada, to Australia, to the United States, to New Zealand.

55 **fen becomes farm** *Guardian*, January 31, 2019.

55 **market gunners** From Idylls, "The Fens," quoted on pp. 363–64, in Miller and Skertchly, *The Fenland Past and Present*. It is possible this passage influenced Aldo Leopold's description of daybreak in the crane marsh in his essay "Marshland Elegy," *A Sand County Almanac*, 1949, 1972, p. 95.

56 **pingo** *Pingo* is the Arctic Inuvialuktun word for a small hill, a desirable hunting feature in flat country.

57 **waterscapes** Laurent Félix-Faure, *Land of Skies and Water, Holland Seen Through the Eyes of Its Painters* (Lemniscaat, Rotterdam, 1996). This beautiful book shows how the Dutch landscape, with similarities to the fenlands of East Anglia, was absorbed into the national consciousness.

57 **ephemeral beauty** Miyazawa Kenji, "Fifth Day, Night," *A Future of Ice* (North Point Press, 1989), p. 16.

57 **malaria** It is not entirely clear if malaria reached England at a particular time or if it was endemic as Mary Dobson states in her *Contours of death and disease in early modern England* (Cambridge University Press, 1997).

58 **"mend the light"** Godalmingmuseum.org.uk/index.php?page=tudor -rushlight.

59 **acequias** The closest practice to this in North America is the centuries-old system of guided, tended and shared acequia irrigation

water in New Mexico villages, a work now fading but immortalized in William deBuys's poetic *River of Traps* (University of New Mexico Press, 1990).

60 **lazy people** Skertchly, *op. cit.*, p. 301. Sir William Dugdale (1605–1686) was the author of *The History of Imbanking and Drayning of Divers Fenns and Marshes, both in foreign parts and in this Kingdom, and of the Improvements thereby extracted from Records, manuscripts, and other authentick testimonies*, 2nd ed., 1772.

60 **unhealthy fens** Dobson, *op. cit.*, pp. 287–92.

60 **"noisome vapors"** *Ibid.*

61 **nuclear testing** Ellen Meloy, *The Last Cheater's Waltz* (Henry Holt, 1999), p. 29.

62 **Waterland** The novel *Waterland* by Graham Swift, 1983, contains rich descriptions of the work of lock-keepers, sluices, dredges, drains, breaks, floods and willful rivers in the nineteenth- and twentieth-century fenlands.

63 **powte** The "powte" was a sea lamprey. Ian D. Rotherham, *The Lost Fens: England's Greatest Ecological Disaster* (History Press, 2013), p. 126.

63 **"a thousand stinks"** Dobson, *op. cit*, pp. 10–11.

64 **malarial fever** Clifford W. Collinson, *Life and Laughter 'Midst the Cannibals* (London, 1926), p. 68.

64 **coffins** Ian D. Rotherham, *op. cit.*, p. 32.

65 **vulnerable women** P. Reiter, "From Shakespeare to Defoe: Malaria in England in the Little Ice Age," *Emerging Infectious Diseases* 6(1) (2000):1–11; Frank Key, "Daniel Defoe and the Fogwives of Essex," thedabbler.co.uk/2015/05/.

65 **marsh fever** Rotherham, *Lost Fens*, p. 36.

65 **a royal brain** *Ibid.*

65 **rare drug** *Ibid.*, p. 43.

66 **the ague** Miller and Skertchly's magnum opus, *The Fenland Past and Present*, devoted an entire chapter to "The Sanitary Condition of the Fens." They approached the subject of "Ague, an endemic, incommunicable, paroxysmal fever" as scientists, looking for causes. They dismissed humidity, remarking, ". . . it is certain that miasmatic emanation in the air is the essential factor in producing ague." They promptly called it "malaria" and that the source was "undoubtedly wet decaying organic matter." They considered the possibility that drinking marsh water could also be a cause, but then differentiated between fen and marsh, finding the acid fens did not give off the bad

air that carried infection. They were sure it was the *rotting* vegetation
of the marshes and went on to discuss at length the varying symptoms
of malaria and the medicines that treat it (quinine) and palliatives that
dull the pain (chiefly alcohol and opium). After ten pages the scientists
went on to ozone, pure water, the water supply and useful statistics
of sewage and gas works. Even as these learned men were writing
up their voluminous reports on the fen world a half-dozen others
were unraveling pieces of the mystery of malaria infection. Nineteen
years after publication of *The Fenland Past and Present*, Ronald Ross, a
doctor in the Indian Medical Service, worked out the bird-mosquito
cycle of malaria infection.

66 **goose and common** For the origin of this witty poem inspired by
English enclosure, see www.cs.ucdavis.edu/~rogaway/classes/188
/materials/boyle.html.

67 **word misuse** It is disturbing to notice that many bloggers and
talking heads are perverting the word *ecosystem* to mean a political
milieu.

68 **bird habitat** www.waxwingeco.com/birding-hotspot.php?id=L1263156;
www.fensforthefuture.org.uk/news/post/cranes.

68 **long-term project** www.greatfen.org.uk/.

69 **return of red admiral** Steven Morris, "Dover Clifftops 'Buzzing
with Wildlife After National Trust Takeover," *Guardian*, July 4, 2020.

70 **impossible project** Bruce Finley, "Booming Front Range Cities Take
First Steps to Build $500 Million Dam, Reservoir near Holy Cross
Wilderness," *Denver Post*, September 6, 2020.

70 **oil under the marshes** www.csis.org/podcasts/babel-translating
-middle-east/azzam-alwash-restoring-iraqs-marshes.

3. BOGS

75 **transformative water** Alan L. Mackay, *The Harvest of a Quiet Eye*
(Institute of Physics, 1977), p. 34.

75 **supernatural quality** An artist interested in "the phenomenality of
water"—not as land- or seascape, but as a substance—is the Irish artist
Remco De Fouw of the Blackstairs Mountains in County Carlow,
whose work with wetness and bogs explores water's qualities.

75 **river water** Norman Maclean, *A River Runs Through It and Other
Stories* (University of Chicago Press, 1974), p. 106.

76 **bird languages** Gilbert White, in Graeme Gibson, *The Bedside Book of Birds* (Doubleday, 2005), p. vii.

76 **confusing peatland words** For a rich discussion of wetland terms and definitions see Hans Joosten, Franziska Tanneberger and Asbjorn Moen, eds., *Mires and Peatlands of Europe: Status, Distribution and Conservation*, Ch. 3, "Mire and Peatland Terms and Definitions in Europe" (Schweizerbart Science Publishers, Stuttgart, 2017), pp. 65–96.

76 **moor and mór** Joosten et al., *Mires and Peatlands of Europe, op. cit.*, p. 72.

76 **heathland** Robert Louis Stevenson, *Kidnapped*, 1886 (Penguin, 1994), p. 154.

77 *meskag–kwedemos* Adrienne Mayor, *Fossil Legends of the First Americans* (Princeton University Press, 2005), p. 12. Although Mayor had ascribed the discovery to the Abenaki, the enthusiast blogger John J. McKay, in writing a book about the mastodon came across historical documents listing the "Indian allies by nation. . . . 237 Iroquois, 50 Abenaki, and 32 Algonquin and Nippising." McKay found that about ninety of the Indian allies deserted near Oswego, New York, "seduced by English brandy." Not many Abenaki were left. Some Shawnees may have been recruited from their settlement near the Scioto River and these people, familiar with the Big Bone Lick likely found the fossils McKay thinks. johnmckay.blogspot.com/2014/01/baron-longueuil-and-mastodon-of-1739.html. Archaeologists think the ancestors of the Shawnee may have been the people who hunted the last mastodons, their Clovis points "unequivocally associated with mastodon remains." See K. Tankersley, M. Waters, and T. Stafford, "Clovis and the American Mastodon at Big Bone Lick, Kentucky," *American Antiquity* 74, no. 3 (July 2009), p. 565.

79 **American paleontology** Elizabeth Kolbert, "The Lost World," *The New Yorker*, December 9, 2013. Cuvier's maiden lecture at the Institute of Science and Arts in Paris in 1796 differentiated between Asian and African elephants and, based on his study of fossils, including those from the "Big Bone Lick," he proposed that in the past monstrous animals had walked the earth and then had disappeared because of some kind of "catastrophe."

79 **racism and misogyny** Many accounts ignore the Indian finders and attribute the discovery to Longueuil. Mayor makes the point that the indigenous finders repeatedly were not credited as the discoverers of these important fossils and she cites the white-man-expedition-

leader superior attitude and dismissive language of the influential paleontologist Gaylord Simpson (1902–1984) as racist. There is, unhappily, a scarlet racist and misogynistic thread that runs through early archaeology and paleontology. Indigenous people, people of color and women who made fossil discoveries were rarely recognized, as in the example of George McJunkin, the Black cowboy who in 1908 discovered the important Folsom site and was not credited with the discovery until recent decades.

80 **western raised bog** The exciting find, by the ecologist Joe Rocchio, is described in the newsletter of the Washington Native Plant Society bulletin for October 29, 2019 (wnps.org/blog/crowberry -musings). The raised bog was "something that had never before been documented in the western United States." The site is now closed to all except a few scientists.

81 **hasty planting** Sharon Levy, "Scotland's Bogs Reveal a Secret Paradise for Birds and Beetles," *Guardian*, November 27, 2019 (originally in I 11.12.2019), p. 3.

81 **opposition** Sharon Levy, *Ibid*.

82 **Lake Nyos deaths** *Ibid*. Some will remember the Lake Nyos outburst of CO_2 gas that killed more than 1,700 people in 1986.

82 **continuing anger** Patrick Barkham, "Row over UK Tree-Planting Drive: 'We Want the Right Trees in the Right Place,'" *Guardian*, February 23, 2021.

82 **methane** Jonathan Watts, "Arctic Methane Deposits 'Starting to Release,' Scientists Say," *Guardian*, October 27, 2020.

82 **deliberate flooding** Cited in Ann Jensen Adams's "Seventeenth Century Dutch Landscape Painting," in *Landscape and Power*, ed. W. J. T. Mitchell (University of Chicago Press, 1994), p. 41.

83 **Vasyugan mire** *Mire* is the umbrella word used in Europe for fens, bogs and swamps.

83 **Cuvette Centrale** G. C. Dargie et al., "Congo Basin Peatlands: Threats and Conservation Priorities, Mitigation and Adaptation Strategies for Global Change" (2018).

84 **sphagnums** E. Thompson, "Natural Communities of Yellow Bogs in Lewis, Bloomfield and Brunswick, Vermont," Technical Report 14, 1989 (Nongame Natural Heritage Program, Vermont Fish and Wildlife Dept.).

84 **James Dog** *OED*, vol. II, p. 358. James Dog was the keeper of Queen Margaret Tudor's wardrobe at the court of Margaret and James IV of Scotland, where the poet William Dunbar was also employed.

84 **modern version** www.wikizero.com/en/Of_James_Dog.
"When I speak to him friendly-like
He barks like a common tyke
[that] chases cattle through a bog.
Madam, you have a dangerous dog."

84 **dangerous bogs** Arthur Conan Doyle, *The Complete Sherlock Holmes*, Vol. I (New York, 2003), p. 623.

85 **escape from the bog** Vladimir Nabokov, *The Stories of Vladimir Nabokov* (Vintage International, 1995), p. 299.

86 **nature photography** Who can forget Samantha Stephens's photo of two little spotted salamanders trapped in a pitcher plant? www .theguardian.com/environment/gallery/2020/dec/24/nature -photographer-of-the-year-2020-the-winners. New technology in infrared cameras shows that pitcher plants can feed at night by luring prey to their luminous glow. Infrared light is invisible to humans but not to insects. The entrapment is shown in the tropical forest episode of the Netflix documentary *Night on Earth*.

87 **dark bogs** John R. Stilgoe, *Common Landscape of America, 1580 to 1845* (Yale University Press, 1982), p. 11.

87 **extinctions** Patrick Greenfield, "Humans Exploiting and Destroying Nature on Unprecedented Scale," *Guardian*, September 10, 2020.

88 **entangled wetland terms** Hans Joosten, Franziska Tanneberger and Asbjorn Moen, eds., *Mires and Peatlands of Europe, op. cit.*

88 **emerging new wetlands** *Ibid.*, p. 65.

88 **Italy for artists** *Ibid.*, p. 8. After Dürer's trip it became obligatory for generations of artists to make the journey to Italy and see how it was done.

89 **"I am a bum"** Robert Hughes, *Goya* (Vintage, 2004), p. 35.

89 **Lesquereux** William C. Darrah, "Leo Lesquereux," *Botanical Museum Leaflets* 2, no. 10 (Harvard University,1934), pp. 113–19, www.jstor .org/stable/41762583.

90 **a brutal doctor** J. P. Lesley, "Obituary Notice of Leo Lesquereux," *Proceedings of the American Philosophical Society* 28, no. 132 (1890), p. 66.

90 **lured to America** Lesley, "Obituary," pp. 65–70. (His name is pronounced "le crew.") Many of Lesquereux's letters have survived. See *Correspondence of Leo Lesquereux and G. W. Clinton*, ed. P. M. Eckel, *Res Botanica* (Missouri Botanical Garden), p. 33.

91 **Lake Drummond a raised bog** John V. Dennis, *The Great Cypress Swamps* (Louisiana State University Press, 1988), p. 50.

91 **lost specimens** Darrah, *op. cit.*, p. 116.

91 **sphagnum virtues** Robin Wall Kimmerer, *Gathering Moss: A Natural and Cultural History of Mosses* (Oregon State University Press, 2003), p. 112.

92 **exploding spores** Dwight L. Whitaker and Joan Edwards, "Sphagnum Moss Disperses Spores with Vortex Rings," *Science* 329 (July 23, 2010), p. 406.

92 **long distances** J. M. Glime, "Adaptive Strategies: Travelling the Distance to Success," Ch. 4–8 in *Bryophyte Ecology*, Vol. 1: *Physiological Ecology*. Ebook sponsored by Michigan Technological University and the International Association of Bryologists. Last updated March 31, 2017, and available at digitalcommons.mtu.edu/bryophyte-ecology/.

92 **noxious gases** Carolyn Kormann, "Annals of a Warming Planet," *The New Yorker*, June 27, 2020.

94 **"sphagnan"** T. Stalheim et al., "Sphagnan: A Pectin-Like Polymer Isolated from Sphagnum Moss Can Inhibit the Growth of Some Typical Food Spoilage and Food Poisoning Bacteria by Lowering the pH," *Journal of Applied Microbiology* 106(3).

94 **footprint** John and Bryony Coles, *People of the Wetlands Bogs: Bodies and Lake Dwellers* (Thames and Hudson, 1989), p. 152.

94 **faceprint** Félix-Antoine Savard, *L'Abatis* (Fides, Montreal, 1934), p. 92.

95 **moss bandage** Flann O'Brien, *At Swim-Two-Birds* (Signet Classic, 1976), p. 179.

96 **bog finds** Wijnand van der Sanden, *Through Nature to Eternity: The Bog Bodies of Northwest Europe* (Amsterdam, 1996), p. 170.

96 **votive offerings** *Ibid.*, p. 175.

97 **absent presence** Franz Lidz, "How the World's Oldest Wooden Sculpture Is Reshaping Prehistory," *New York Times*, www.nytimes.com/2021/03/22/science/archaeology-shigir-idol.html.

97 **Shigir Idol** *Ibid.*

99 **a famous ending** Frank O'Connor, *Collected Stories* (Vintage, 1982), pp. 8, 12.

99 **common findings** Bridget Brennan, "The Influence of Shamanistic Practice on the Deposition of Prehistoric Human Remains in Bogs," thesis, 2014, p. 2. Wijnand van der Sanden points out the "triple death motif" found in medieval Irish and Welsh stories; Van der Sanden, *Through Nature to Eternity*, *op. cit.*, p. 175.

99 **shamanism** *Ibid.*

100 **"ur-camera"** Karin Sanders, *Bodies in the Bog and the Archaeological Imagination* (University of Chicago Press, 2009).

100 **dissolved bone** Van der Sanden, *Through Nature to Eternity*, *op. cit.*, p. 16.

102 **emotional responses** W. Somerset Maugham, *The Complete Short Stories*, Vol. I (East and West, NY, 1952), pp. 131–32.

103 **Josef Beuys** Sanders, *op. cit.*, 126

103 **Bachelard** Bachelard, *Earth and Reveries of Will*, p. 8, quoted in Sanders, *op. cit.*, p. 133, n. 264.

104 **Naboland** Peter Davidson, *The Idea of North* (Reaktion Books, London, 2005, 2007), pp. 109 ff.

104 **peat preservation** Robert Van de Noort and Aidan O'Sullivan, *Rethinking Wetland Archaeology* (Duckworth, London, 2006).

104 **Fayoum portraits** Alexxa Gotthardt, "Unraveling the Mysteries of Ancient Egypt's Spellbinding Mummy Portraits," Artsy, www.artsy.net.

105 **lur** To hear the lur go to www.youtube.com/watch?v=Ld6Dt -Lce6M.

106 **digging up the devil** Van der Sanden, *Through Nature to Eternity*, *op. cit.* (Amsterdam, 1996), p. 37.

106 **Alfred Dieck** *Ibid.*, 54.

107 **choosing stories** Randall Jarrell, Introduction to *The Anchor Book of Stories* (Doubleday Anchor, 1958), p. xv.

108 *infames corpores* Tacitus, "The Germany and the Agricola of Tacitus," www.gutenberg.org/files/7524/7524-h/7524-h.htm.

108 **Himmler speech** Sanders, *op. cit.*, pp. 61–62.

108 **Sanders** www.loebclassics.com/view/livy-history_rome_22/2019/ pb_LCL233.261.xml. Karin Sanders, *Bodies*, links the meaning of *corpores infames* as "homosexual" to Himmler (p. 62).

109 **slaves were non-persons** Philippe Ariès and Georges Duby, eds., *A History of Private Life*, Vol. 1: *From Pagan Rome to Byzantium* (Harvard University Press, 1987), p. 204.

110 **Windeby "girl"** Seamus Heaney, *North* (Faber and Faber, 1975, 1992), pp. 30–31.

110 **Irish Troubles** Navleen Multani, "Bog Body, Violence and Silence in Seamus Heaney's 'Punishment,'" *Dialog* 34 (2019).

111 **Windeby Child** Jarrett A. Lobell and Samir S. Patel, "Windeby Girl and Weerdinge Couple," *Archaelogy* 63, no. 3 (2010).

111 **a failed king?** Kristen C. French, "The Curious Case of the Bog Bodies," https://nautil.us/issue/27/dark-matter/the-curious-case-of -the-bog-bodies.

112 **post-ice people** Steven Mithen, *After the Ice: A Global Human History 20,000–5000 B.C.* (Weidenfeld & Nicolson).

112 **importance of personal experience** *Ibid.*, p. 116.

113 **McJunkin's Clovis point** George McJunkin, an ex-slave turned cowboy who swapped lessons in horse-breaking for lessons on how to read, first found the famous site in 1908. Aside from two amateur fossil-hunters he could not get any archaeologist or institution interested in his find of giant bison bones with a stone projectile embedded in one. After his death in 1922 the fossil hunters kept at it and finally claimed the important discovery. It is still not McJunkin's name that you read in Clovis point discovery accounts.

114 **stellar preservation** Mithen, *op. cit.*, p. 231.

114 **White Sands footprints** Carl Zimmer, "Ancient Footprints Push Back Date of Human Arrival in the Americas," *New York Times*, September 23, 2021, https://www.nytimes.com/2021/09/23/.

116 **short Romans** Kyle Harper, "The Environmental Fall of the Roman Empire," *Daedalus*, Spring 2016, p. 103.

116 **Wingfield sleeps** Tobias Wolff, *In the Garden of the North American Martyrs* (HarperCollins, 1981).

117 **invention of Teutoburg Forest** A seventeenth-century church pastor had changed the name of woodland near Detmold, "Lippe-Raum," to "Teutoburg Forest" for reasons of his own and it was thus marked on the diocese map.

117 **"Herman the German"** A few years later a monument to Arminius/ Hermann was put up in New Ulm, Minnesota, aka "Herman the German," but that's another story.

117 **narrow pass** Tony Clunn, *The Quest for the Lost Roman Legions* (Savas Beatie, NY, 2005), p. 82.

117 **old hunting strategy** Mithen, *op. cit.*, p. 122.

118 **a silver coin** *Ibid.*, p. 4. Clunn also included a discussion of the scholars' acceptance of "saltus" as a gap rather than a forest.

119 **lead slingshot proof** *Ibid.*, p. 26.

120 **Milius-MacDonald-Heller *Rome*** History buffs who watched John Milius's, William J. MacDonald's and Bruno Heller's brilliant television series *Rome*, released in 2005, are familiar with the young Octavian, Caesar's heir, who became Rome's first emperor, the wily Augustus (27 BC–AD 14).

120 **how tribal "leaders" emerge** Peter S. Wells, *The Barbarians Speak: How the Conquered Peoples Shaped Roman Europe* (Princeton University Press,

1999), pp. 116 ff. Jonathan Hill, *History, Power and Identity: Ethnogenesis in the Americas 1492-1992* (University of Iowa Press, 1996).

121 **Varus as Custer** Simon Schama, *Landscape and Memory*, p. 88.

121 **Arminius as Roman auxiliary** Clunn, *op. cit.*, p. 26.

122 **a warning** Wells, *op.cit.*, p. 40.

4. SWAMP

128 **ice age melt** Its torrential freshwater discharge into the Arctic Ocean may have shifted ocean currents and perhaps been a factor in starting up the bitter Younger Dryas period. There is more Big Water we didn't know about. Evidence from ancient ocean crust sampling is beginning to build up a theory that the oceans were twice as large in the past as they are now, and that land was in scarce supply 3–4 billion years ago. There is also much water in ringwoodite, one of the earth's transitional mantle layers.

128 **observation acknowledged** John Soennichsen, *Washington's Channeled Scablands Guide* (Seattle, 2012), pp. 21 ff. "Bretz was widely criticized by his contemporaries for his 'outrageous hypothesis' while continuing to use his own observations, not popular opinion, to develop his ideas." www.historylink.org/File/8382.

129 **intimacy of Mesolithic tools** Steve Mithen, *op. cit.*, pp. 304 ff. Mithen expresses a sense of this oneness with the natural world in his discussion of Mesolithic microlith tools: "The delight of such tools is that they appear to seep from nature itself; they tell of an intimacy with the natural world that is lost today and are the handiwork of people who loved their craft."

129 **population almost doubles** Thomas E. Dahl and Gregory J. Allord, "History of Wetlands in the Coterminous United States," National Water Summary on Wetland Resources, U. S. Geological Survey Water Supply Paper 2425, 1996, pp. 3–4.

129 **war is mud** Weymouth T. Jordan, "'Drinking Pulverized Snakes and Lizards': Yankees and Rebels in Battle at Gum Swamp," *The North Carolina Historical Review* 71, no. 2, 1994, pp. 207–31.

130 **not a river but a swamp** Dante Alighieri, *The Divine Comedy*, trans. Lawrence Grant White (Pantheon, 1948), p. 13. *The Divine Comedy of Dante Alighieri. The Italian Text with a Translation in English Blank Verse and a Commentary by Courtney Langdon, vol. 1 (Inferno)* (Cambridge:

Harvard University Press, 1918). The Italian original used in the Langdon translation is:

L'acqua era buia assai più che persa; / e noi, in compagnia dell' onde bige, / entrammo giù per una via diversa. / Una palude fa, che ha nome Stige, / questo tristo ruscel, quando è disceso / al piè delle maligne piagge grige. (There is no resolution, as *palude* can be translated as "swamp" *or* "marsh.")

131 **salt marsh words** John R. Stilgoe, *Shallow Water Dictionary* (Princeton Architectural Press, 1994).

134 **dream house** Henry D. Thoreau, "Walking," 1861.

134 **the politics of wetlands** Thomas E. Dahl and G. J. Allord, "Wetlands Losses in the United States 1780s to 1980s," epa.gov./water/archive. For a scorching commentary on George Bush's and Quayle's broken political promises see Mark Monmonier, *Drawing the Line* (New York, 1990), p. 30.

134 **Hampton Roads** "Hampton Roads" is also the terrestrial region of nine cities around Chesapeake Bay.

135 **old swamps want to return** Jeff Goodell, *The Water Will Come*, 2017, p. 192.

135 **man—disturbing agent** George P. Marsh, *The Earth as Modified by Human Action* (New York, 1874), p. 34.

136 **killing ground** Michael Grunwald, *The Swamp, the Everglades, Florida, and the Politics of Paradise* (New York, 2006, 2007), p. 121.

136 **market hunters** Frank G. Ashbrook and Edna Sater, *Cooking Wild Game* (New York, 1945), pp. 3–4. Included were recipes for "Possum with Tomato Sauce . . . Muskrat Meatloaf . . . Fried Coot." Today's equivalent cartoonist is the Australian Andrew Marlton, whose sarcastic "anarcho-marsupialist" strip *First Dog on the Moon* has a global following.

137 **a new idea** www.thebeatnews.org/BeatTeam/history-federal -wetland-protection/.

138 **book travel with Bartram** William Bartram, *Travels Through North and South Carolina, Georgia, East and West Florida*, 1790, aka *Travels and Other Writings* (Library of America, 1976).

138 **Franklinia** John Bartram sent plant specimens he had collected to Linnaeus, and helped Linnaeus's student Peter Kalm with his collection when he visited America. Bartram's house was on the outskirts of Philadelphia. The garden exists today as "Bartram's Garden."

138 **discovery** *Travels, op. cit.*, p. 38.

138 **quinine** Richard L. Thornton, "American Farmers Could Be Growing the Tree for Producing Quinine Right Now!," The Americas Revealed, March 29, 2020, apalacheresearch.com/2020 /03/29/.

138 **choice game** Bartram, *Travels*, p. 103.

139 **kept up by owls** *Ibid.*, p. 126. Courting owls can make a tremendous racket through the spring nights.

139 **swamp succession** Meanley, *The Patuxent River Wildrice Marsh*, 1993, p. 5.

141 **other predators** Meanley, *Blackbirds of the Southern Rice Crop* (U.S. Dept. of the Interior, Fish and Wildlife Service, Resource Publication 100, 1971), p. 30.

141 **20 million birds** Meanley, *Swamps*, p. 95. "Slovakia" was settled by immigrant Slavs in the 1890s.

142 **rich wildlife in Singer Tract** Meanley, *op. cit.*, p. 90.

143 **extinct** Brooks Hays, "Fish and Wildlife Service announces extinction of 23 species in 19 states," UPI, September 29, 2021. www .upi.com/Science_News/2021/09/29/.

143 **Okefenokee has everything** Meanley, *op. cit.*, 13.

144 **Meanley's observation in 1972** *Ibid.*

144 **cypress knees** Also suggesting the British music hall and dance hit before World War I—"Knees Up, Mother Brown."

145 **burning the trees** B. W. Wells and L. A. Whitford, "History of Stream-Head Swamp Forests, Pocosins, and Savannahs in the Southeast," *Journal of the Elisha Mitchell Scientific Society* 92, no. 4, 1976, pp. 148–50.

145 **importance of pocosins** oceanservice.noaa.gov/facts/pocosin.html.

146 **"haunted by waters"** Norman Maclean, *A River Runs Through It and Other Stories* (University of Chicago Press, 1976, 1983).

147 **people as tools** William Byrd and E. G. Swem, *Description of the Dismal Swamp and a Proposal to Drain the Swamp* (Metuchen, NJ: Printed for C. F. Heartman, 1922). PDF retrieved from the Library of Congress, www.loc.gov/item/22022884/, pp. 16–17.

147 **charms of plantation life** William Byrd, quoted in Bland Simpson, *The Great Dismal, op. cit.*, p. 63.

148 **Henry Lee** *Encyclopedia of North Carolina*, ncpedia.org/great-dismal -swamp. Henry Lee, aka "Lighthorse Harry" (1756–1818), was one of the founding fathers. A brilliant horseman and officer of a cavalry unit in the Revolutionary War, he was also an incompetent investor who bungled his money affairs, went bankrupt and to debtor's prison.

He wrote Washington's eulogy "First in war, first in peace, first in the hearts of his countrymen." Supporting his editor friend Alexander Hanson, who published the *Federal Republican*, he was beaten and maimed and spent years abroad before returning to die in 1818. His CSA commander son, Robert E. Lee, is better known.

149 **expert of southern swamps** Brooke Meanley, *The Great Dismal Swamp* (Audubon Naturalist Society of the Central Atlantic States, 1973). Bland Simpson, when researching his account of the Dismal, consulted and met with Brooke Meanley, joining him and his wife, Anna, in the swamp for bird walks.

149 **bird millions** Brooke Meanley, *Swamps, River Bottoms & Canebrakes* (Barre, MA, 1972), p. 38.

150 **walking through a tree** Bland Simpson, *Dismal, op. cit.*, p. 29.

150 **clear-cuts** John V. Dennis, *The Great Cypress Swamps* (Louisiana State University, 1988), p. 50.

153 **neighbors in the way** Ben Palmer, *Swamp Land Drainage with Special Reference to Minnesota* (University of Minnesota Bulletin, 1915), p. 32.

154 **Lake Erie pollution** Sharon Levy, "Learning to Love the Great Black Swamp," undark.org/2017/03/31/.

155 **nothing left** Mitsch and Gosselink, *op. cit.*, p. 482. There is a video of the Kankakee at www.ben-hur.com/everglades-of-the-north.

155 **a very great loss** Andrea Neal, "Draining of 'Everglades of the North' Destroyed a Hoosier Habitat," kpcnews, January 18, 2016.

156 **descriptive words** Charles H. Bartlett, *Tales of Kankakee Land*, 1904, pp. 24–25.

157 **who got the drained land?** Jack Klasey, "Looking Back: Murdering the Grand Kankakee Marsh," February 23, 2019, www.daily-journal .com.

158 **difficult restoration** www.reconnectwithnature.org/News-Events/ News/State-Grant-Awarded-Braidwood-Dunes-Kankakee-Sands.

159 **the crane** Aldo Leopold, *A Sand County Almanac*, p. 96.

159 **nature's "gifts"** Curt Burnette, "Limberlost, Kankakee, Black Swamp," *Friends of the Limberlost*, January 24, 2019.

161 **shape of mangrove "skirts"** Whale baleen, cane, brass strips and, finally, flattened steel wire supported the crinolines and fabrics of this uncomfortable fashion.

162 **inside a mangrove forest** Mitsch, *op. cit.*, p. 311. *Impenetrable* is the key word. Mangroves invite comparison with the spiky nightmare of invasive Himalayan blackberry in the Pacific Northwest, or the tortured interwoven branches of Newfoundland-Labrador tuckamore.

163 **oil refinery** Madelin Andersen, "Loss of Mexico's Valuable Mangrove Forests"; Carrie Madren, "Mangroves in the Mist," *American Forests*; Greg Allen, "Climate Change May Wipe Out Large Mangrove Forests, New Research Suggests," *All Things Considered*, September 11, 2020; Jordan Davidson, "Mexico Is Letting an Oil Company Destroy Protected Mangroves for an $8 Billion Oil Refinery," EcoWatch, March 6, 2020.

164 **Lewis an observer** Thoreau was a passionate and intense observer using all his senses. His method is described in Richard Higgins, *Thoreau and the Language of Trees* (University of California Press, n.d.), p. 10 ff.

165 **understanding mangroves** Hannah Waters, "Mangrove Restoration: Letting Mother Nature Do the Work," ocean.si.edu/ocean-life/plants-algae/mangrove-restoration-letting-mother-nature-do-work.

165 **secret of success** *Ibid.*

166 **CO_2** In September 2018 World Heritage and ICUN forestry experts began an examination of the forest to review then-current management practices. The report was to be discussed in 2019 at the World Heritage Committee annual session in Baku. whc.unesco.org/en/news/1884/.

167 **Pantanal-Vasyugan size** Sergey Kirpotin et al., "Great Vasyugan Mire: How the World's Largest Peatland Helps Address the World's Largest Problems," *Ambio*, 2021.

168 **Vasyugan's sinuous curve** globolo.ru/en/vasyuganskoe-boloto-na-karte-vasyuganskie-bolota-rossiya-socialnoe-i.html. This information source is in garbled English and it is difficult to understand meanings. We read: "The people go to the legend of the origin of Vasyugan swamps. It turns out that the swamps created the hell himself, he created the land, dismissed with water with thickets of coarse herbs and curvature trees. The legend says that at first there was no sushi on the ground, there was only water and God went around her. One day, he saw a muddy bubble . . ." It reads like an egregiously ham-fisted AI "translation" although "sushi" is a nice touch.

168 **last refuge** Kirpotin et al., "Great Vasyugan Mire," *op. cit.*

168 **rare slender-billed curlew** magornitho.org/2019/04/.

169 **carbon bomb?** *Ibid.*

170 **prairie potholes** Mitsch and Gosselink, *Wetlands*, 5th ed., 2015, pp. 61–62.

171 **unhappy choice** Warren Cornwall, "Unleashing Big Muddy," *Science*, April 23, 2021, p. 334.